The New World of Utilities

Vincent Petit

The New World of Utilities

A Historical Transition Towards a New
Energy System

 Springer

Vincent Petit
Schneider Electric
Hong Kong

ISBN 978-3-030-00186-5   ISBN 978-3-030-00187-2   (eBook)
https://doi.org/10.1007/978-3-030-00187-2

Library of Congress Control Number: 2018956562

This Springer imprint is published by the registered company Springer Nature Switzerland AG
The registered company address is: Gewerbestrasse 11, 6330 Cham, Switzerland

List of Countries

Asia (Non-OECD)
Samoa, Bangladesh, Brunei Darussalam, Bhutan, Channel Islands, Fiji, Guam, Micronesia, Fed. Sts., Indonesia, Cambodia, Kiribati, Korea, Rep., Lao PDR, Sri Lanka, Maldives, Marshall Islands, Myanmar, Mongolia, Northern Mariana Islands, Malaysia, Nepal, Pakistan, the Philippines, Palau, Papua New Guinea, Singapore, Solomon Islands, Suriname, Timor-Leste, Thailand, Tuvalu, Taiwan, China, Vietnam, Vanuatu

Africa
Aruba, Angola, Burundi, Benin, Burkina Faso, Botswana, Central African Republic, Cote d'Ivoire, Cameroon, Congo, Rep., Comoros, Cabo Verde, Djibouti, Algeria, Egypt, Arab Rep., Eritrea, Ethiopia, Gabon, Ghana, Guinea, Gambia, Guinea-Bissau, Equatorial Guinea, Kenya, Liberia, Libya, Lesotho, Morocco, Madagascar, Mali, Mozambique, Mauritania, Mauritius, Malawi, Namibia, Niger, Nigeria, Rwanda, São Tomé and Príncipe, Sudan, Senegal, Sierra Leone, Somalia, South Sudan, Swaziland, Chad, Togo, Tonga, Tunisia, Tanzania, Uganda, South Africa, Congo, Dem. Rep., Zambia, Zimbabwe

Eurasia
Afghanistan, Armenia, Azerbaijan, Belarus, Georgia, Kazakhstan, Kyrgyz Republic, Russian Federation, Tajikistan, Turkmenistan, Uzbekistan

Europe
Andorra, Albania, Austria, Belgium, Bulgaria, Bosnia and Herzegovina, Central Europe and the Baltics, Switzerland, Cyprus, Czech Republic, Germany, Denmark, Spain, Estonia, Finland, France, Faeroe Islands, United Kingdom, Greece, Grenada, Greenland, Croatia, Hungary, Isle of Man, Ireland, Iceland, Italy, Kosovo, Liechtenstein, Lithuania, Luxembourg, Latvia, Monaco, Moldova, Macedonia, FYR, Malta, Montenegro, Netherlands, Norway, Poland, Portugal, Romania, Serbia, Slovak Republic, Slovenia, Sweden, Ukraine

Middle East
United Arab Emirates, Bahrain, Iran, Islamic Rep., Iraq, Israel, Jordan, Kuwait, Lebanon, Oman, Qatar, Saudi Arabia, Syrian Arab Republic, Turkey, West Bank and Gaza, Yemen Rep.

North America
Canada, Mexico, United States

OECD Asia
Australia, Japan, New Caledonia, New Zealand, Korea, Dem. Rep., Pacific island small states, French Polynesia

South America
Argentina, Antigua and Barbuda, Bahamas, The, Belize, Bermuda, Bolivia, Brazil, Barbados, Chile, Colombia, Costa Rica, Caribbean small states, Cuba, Curacao, Cayman Islands, Dominica, Dominican Republic, Ecuador, Guatemala, Guyana, Honduras, Haiti, Jamaica, St. Kitts and Nevis, St. Lucia, St. Martin (French part), Nicaragua, Panama, Peru, Puerto Rico, Paraguay, El Salvador, Sint Maarten (Dutch part), Seychelles, Turks and Caicos Islands, Trinidad and Tobago, Uruguay, St. Vincent and the Grenadines, Venezuela, Virgin Islands (U.S.)

Contents

List of Figures

List of Tables

Chapter 1
Introduction

Electricity is a form of energy which has long been known. Early mention can be found in numerous texts from ancient Egypt which covered the power of electric eels. The effects of electricity started to be studied only at the beginning of the seventeenth century, by the English scientist William Gilbert, who actually gave electricity its name from the Greek word "elektron", which meant amber. He indeed identified the effects of static electricity produced by rubbing amber. Later on, during the eighteenth century, Benjamin Franklin devoted significant resources to the study of static electricity. However, it was only in the late nineteenth century that this new form of energy took off as an energy source for industry. The first electrical distribution system was built by Thomas Edison in 1882 in New York near Pearl Street Station and distributed electricity to a small area of lower Manhattan (Institute for Energy Research 2014). A number of small systems then started to spread across the United States and Europe. They ran in isolation and provided electric power to factories and mansions. The development of alternative current (AC) eventually enabled transmission of power over long distances and interconnection between various sources of supply, making it possible to widely distribute electricity to households and consumers. Power distribution then started to expand throughout the world, propelling the second industrial revolution, and electricity progressively became a major source of energy supply.

Today electricity represents almost 30% of the total consumption of energy in buildings, and more than 25% of the energy consumption in industry (© OECD/IEA, WEO 2014). Electricity is used to provide lighting and heating in homes as well as to power electric appliances that populate buildings and homes, such as refrigerators, cooking appliances, TV sets, computers, air conditioning equipment, etc. In industry, it is used to power electric motors, which account for close to 70% of the total electricity consumed in the sector (© OECD/IEA, Motors 2011). Electricity consumption continues to rise. It increased by 87% in the past 20 years and is forecasted to increase by up to another 85% by 2035 (© OECD/IEA, WEO 2014). In comparison, total energy consumption increase was "only" 35% in the past 20 years, with growth in the coming 20 years projected to be of similar magnitude. Electricity's

© Springer Nature Switzerland AG 2019
V. Petit, *The New World of Utilities*, https://doi.org/10.1007/978-3-030-00187-2_1

share in the overall energy mix thus keeps increasing as it becomes an everyday source of energy. Various factors account for this increasing share. They include economic development, rising living standards throughout the world, increased urbanization, and the massive potentialities of new digital technologies. The distribution of power has thus never been so important for the economic and social development of the world population.

Developments in distribution of electricity surged in the twentieth century. From small isolated power companies, the market evolved towards a more centralized model that is heavily regulated. Electricity had become a strategic resource, and large companies, in a monopolistic situation, provided an end-to-end service to consumers, from operating generation power plants, controlling transmission and distribution networks, down to the actual supply of the commodity to millions of consumers. This situation changed in the last quarter of the twentieth century. In search of efficiency, market regulators pushed the market towards deregulation. While the distribution of electricity remained regulated, a deregulated market for power generation and electricity supply to consumers emerged in a number of countries. The situation varied considerably across regions and countries. The movement essentially started in the United States and Europe. In other regions of the world, many countries still rely today on heavily regulated and integrated power markets. The situation here has remained very dependent on the market regulator's choices. Nevertheless, over the years, electricity markets became increasingly complex to operate, with the emergence of new utilities which cover part or all of the value chain.

Electricity is produced by many different sources. Steam generators were among the first sources of power. The steam drove the mechanical movement of a turbine which, within an alternator, produced electricity. Nowadays, most power plants use a similar principle. Coal, natural gas, biomass, geothermal energy, and even nuclear fission, provide the needed heat to propel the turbine within an alternator. Renewable energies were also used from the beginning of the electric era. Hydroelectric dams used the energy of flowing water to power turbines. For close to two decades now, other new sources of energy have emerged. Wind turbines use the flow of wind to rotate a turbine coupled to an alternator. Solar farms operate on a different principle. They use the properties of semiconducting materials to directly generate electricity from solar radiation. The development of these energies was first subsidized by governments as a political means to reduce the carbon footprint of electricity production. Indeed, 68% of the electricity generation in 2012 came from power plants using fossil fuels, and the associated carbon emissions represent today around 35% of overall emissions, according to the Intergovernmental Panel on Climate Change (2007). The massive deployment of wind and solar generation led to a strong drop in prices over the past years. Prices of solar photovoltaic modules have for instance been divided by five over the 2008–2013 period, following the well-known learning curve of mass production (© OECD/IEA, Solar 2014). The cost of electricity supplied by renewable sources is now becoming increasingly competitive in all regions of the world against traditional sources of power. Forecasts from various sources actually confirm the trend towards a further drop of these costs in the coming decades, leading to a massive penetration of renewable electricity into the market. This continuous drop of prices yields a new paradigm, one in which renewable electricity could become the cheapest source of energy across the entire energy system. Now, these new sources of energy create a number of challenges as their shares of the overall power generation mix grow. First, their actual marginal cost is null, leading to important evolutions of price levels at certain times of the day, when they operate at full power, destabilizing traditional wholesale power markets. Second, their inherent intermittency (the result of dependence on weather conditions) creates a number of issues when it comes to balancing real-time supply and demand. It also leads them to be prioritized on the network, as they cannot be regulated like traditional sources of power. This drastically changes the size of the market in which traditional generation sources can operate, at least at certain times of the day. Finally, the traditional architecture of power distribution was designed in the twentieth century on a centralized model, where power generation was connected to a high-voltage network transmitting power throughout a region or a nation, then being distributed to end users at a lower voltage in a one-way flow. The emergence of renewable energies, which are by nature distributed, increases the complexity of distribution as power now flows in a multitude of directions. The emergence of distributed renewable energies is thus considerably changing the face of power distribution, creating a new realm of threats *and* opportunities for market players, as well as complex technical challenges.

The purpose of *The New World of Electric Utilities* is to illustrate these changes at stake, and to show how technology and innovations can help enable this transition towards a new electric world, where the potentialities of this source of energy, which has already contributed greatly to the economic and social development of the world in the twentieth century, could be guaranteed going forward. There are a number of books which already describe these changes, with a variety of scenarios, but few (if not none) provide and discuss quantitative forecasts. As well, there are a number of works on power generation and grid management issues, but few embrace the complete value chain from generation to consumption, although the issues are intertwined. The purpose of this work is thus to provide quantitative details on all of the value chain so the reader can better understand the major changes at stake, as well as the significant uncertainties which remain and which will considerably influence the path forward.

In the first chapter, the evolutions of electricity demand will be reviewed and forecasts on electricity consumption discussed for the coming decades. A specific model will be described to highlight the evolutions of electricity consumption, and the massive potentialities which can influence electricity consumption upwards or downwards. Then, the evolution of power generation mix and, more importantly, competitiveness across power generation sources will be reviewed. Similarly, a specific model will be described to highlight and help project the upcoming evolutions in terms of competitiveness and the power generation mix. Again, the future is not yet written and a number of different evolutions could take place, depending on upcoming decisions. These evolutions will be reviewed. In a third part, the impact of the changes of demand and mix on the overall management of the grid and the market in general will be studied. A quantitative model will be presented to forecast the upcoming evolutions and identify the main challenges at stake. In a final chapter, the key solutions to smooth the transition now in progress will be reviewed.

In the coming decades, the energy landscape will progressively transition towards a more decentralized and connected system. To a large extent, electric energy will also become more affordable (if not free over time), yielding massive productivity step changes. This transition is thus the dawn of a new era where clean, distributed and cheap (if not free) energy will replace polluting, expensive, and non-sustainable resources. It is a drastic change of paradigm for the world population, in particular for those in great need of energy to sustain their rapid development. The future is electric!

References

© OECD/IEA, Motors. (2011). https://www.iea.org/publications/freepublications/publication/EE_for_ElectricSystems.pdf

© OECD/IEA, Solar. (2014). https://www.iea.org/media/freepublications/technologyroadmaps/solar/TechnologyRoadmapSolarPhotovoltaicEnergy_2014edition.pdf

© OECD/IEA, WEO. (2014). In Greenpeace: The energy [R]evolution scenario. http://www.greenpeace.org/international/Global/international/publications/climate/2015/Energy-Revolution-2015-Full.pdf and https://www.iea.org/publications/freepublications/publication/WEO2014.pdf

Institute for Energy Research. (2014). http://instituteforenergyresearch.org/electricity-distribution/

Intergovernmental Panel on Climate Change. (2007). http://www.ipcc.ch/pdf/assessmentreport/ar4/syr/ar4_syr_fr.pdf

Chapter 2
The Transition to a New Electric World

Since the emergence of electricity as a reliable source of energy at the end of the nineteenth century, power distribution has expanded dramatically, representing a significant share of the energy used in buildings and by the industry sector. This trend is projected to continue in the coming decades due to a multitude of factors and opportunities, which could make the world of tomorrow an (almost) all-electric world.

2.1 Spectacular Growth of Electricity Consumption

2.1.1 All Forecasts Confirm Trend Towards a Significant Increase

The overall electricity consumption in the world has almost doubled in the last 20 years, jumping from around 9000TWhin 1990 to about 17,000TWh in 2010 (Fig. 2.1). This increase in electricity consumption parallels global energy consumption, which has arisen by 35% in the past 20 years. Electricity consumption thus grew faster than for overall energy, which means that electricity now has a bigger share of the global energy mix.

China has been a main driver of growth in the past 20 years, posting an increase of almost 3000TWh over the time period, or 37% of the world's total increase in electricity consumption. Altogether, new economies represented two thirds of the growth, with over 5200TWh of increase in consumption. A large part of the electricity consumption increase is thus explained by the growth in new economies. On average, electricity consumption in China increased by 10% every year; the annual increase in the rest of the new economies was 5.5%. The remaining share of the increase in electricity consumption has come from mature economies, which have on average grown by slightly less than 2%, with the exception of Eurasia,

© Springer Nature Switzerland AG 2019
V. Petit, *The New World of Utilities*, https://doi.org/10.1007/978-3-030-00187-2_2

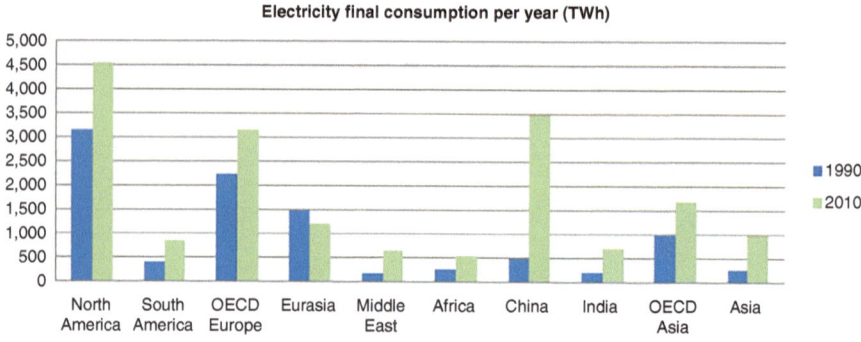

Fig. 2.1 Electricity final consumption (© OECD/IEA, WEO 2012)

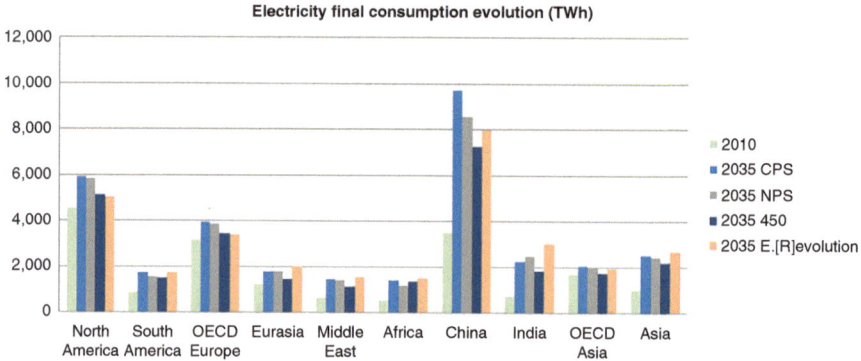

Fig. 2.2 Electricity final consumption to 2035 (Greenpeace 2015; © OECD/IEA, WEO 2014)

which electricity consumption has actually dropped by an average of 0.7% per year in the past 20 years (a detailed list of countries by region is included in the annex).

In the coming 20 years, electricity consumption is projected to increase by another 85% according to forecasts by the International Energy Agency (2012, 2014), and reach over three times its 1990 levels (Fig. 2.2). There are several forecasts which slightly vary from one another, depending on the scenario considered. The International Energy Agency uses a "Current Policy" (CPS) scenario, which forecasts the evolution of energy consumption assuming the continuation of current trends. It also forecasts a "New Policy" (NPS) scenario, which forecasts energy consumption evolution based on a reasonable and expected evolution of the policy framework (towards more energy efficiency, for instance). It studies as well a "450" scenario, which evaluates the energy consumption evolution when governments across the globe take measures to limit the greenhouse gas concentration in the atmosphere to 450 ppm. Finally, the "Energy [R]evolution scenario" by Greenpeace (2015) identifies more radical options in terms of energy supply and policy measures. There exist a number of other scenarios and projections from various other sources which are not being presented here. For instance, Exxon Mobil (Exxon Mobil 2016) and BP (BP 2016) have rather conservative forecasts that are close to or more conservative than the ones from the International Energy Agency (2012). A number of agencies are also publishing forecasts, such as the Ecofys report from WWF (2011) or the Global Energy Assessment from the International Institute of Applied System Analysis (2012). Most of these projections range between the ones used here, at least by 2035, which is why they are not presented here.

In the end, the forecasted trends remain the same. Overall the reference CPS scenario estimates that global electricity consumption in 2035 will reach 33,000TWh (or 85% of growth) whereas the forecasts for the NPS, 450 and Energy [R]evolution scenarios are lower. The Energy [R]evolution scenario forecast is very close to that for the NPS scenario, at around 31,000TWh. It predicts globally higher forecasts in new economies, with lower forecasts for OECD countries and China. The 450 scenario forecasts electricity consumption in 2035 at around 27,000TWh, about 17% lower than the one from the CPS scenario, a result of the deployment of all the energy efficiency measures required to meet its objectives. In any case, in all the projections, the trends towards growth remain the same in every region.

The projections of electrical consumption growth range from 1.7% to 2.5% annualized growth worldwide, depending on the scenarios. In all scenarios, non-OECD countries correspond to 80–90% of the total growth. China alone corresponds to between 35% of the overall growth in the Energy [R]evolution scenario and 41% in the CPS scenario, with electricity consumption expected to more than double in 20 years. The country corresponds in those forecasts to a little less than half of the growth of electricity consumption in non-OECD countries. India would post the highest growth in all scenarios, with an annualized increase in consumption of 6–7.5% as it catches up with China in terms of economic growth. Electricity consumption in India would be multiplied by three to four in the various scenarios reported here in a period of 20 years. Finally, OECD countries would not grow more than 1.4%. In the Energy [R]evolution scenario, the growth in these countries is even lower, around 0.5% on average.

Electricity consumption per capita has also evolved over the past 20 years and is projected to continue to do so in the coming decades (Fig. 2.3). It is calculated based on the CPS scenario and estimates of world population growth (Our World in Data 2016). It has basically increased in all regions of the world, although not at the same speed. The annualized increase worldwide was 1.7%, with several regions consistently above this growth average. China posted electricity consumption growth per capita of nearly 10%; growth in the Middle East, India and Asia was about 5%. Africa posted growth per capita of only 1.5% in the past 20 years, similar to Europe, a clear indication of a lower speed of economic development in the region. The trend in the coming 20 years should be upwards, with India and China growing around 4% per year, Asia growing around 3%, and the Middle East around 2% per year. Africa is expected to continue to lag in the CPS scenario with an annualized growth of electricity consumption per capita of 1.2%, lower than the world average.

In summary, electricity consumption is expected to increase in all regions of the world. Non-OECD countries should experience an annualized growth around 3% in

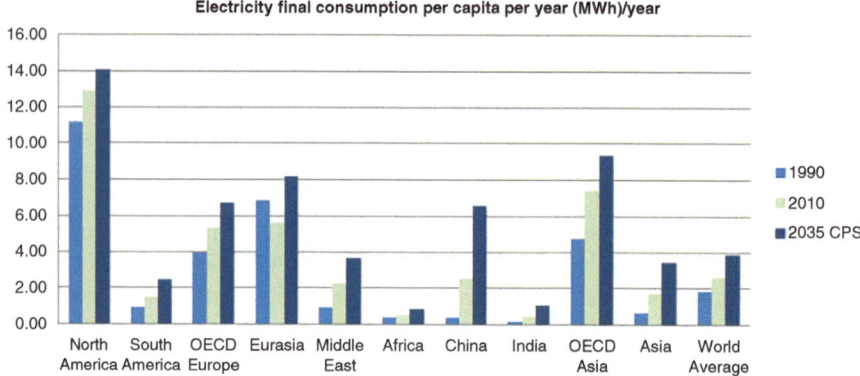

Fig. 2.3 Electricity final consumption per capita (© OECD/IEA, WEO 2012; Our World in Data 2016)

all scenarios, leading many countries to see their consumption more than double in the next 20 years (and being multiplied by close to four in India). The differences of electricity consumption per capita and their projections also suggest that these forecasts may be very conservative as, with the exception of China, there is no change of paradigm in the coming 20 years for most non-OECD countries, which would remain in a situation of "electric poverty".

2.1.2   Increased Usage in All Sectors, but Dynamics Vary

End use electricity consumption varies from one sector to another, with the dynamics different in each. The past 20 years showed accelerated growth in the buildings sector, with consumption doubling, while consumption in the industry sector increased by 70%. Electricity consumption in the transportation sector slightly increased during this period, although its overall size is negligible (around 2% of the total). The buildings sector is slightly bigger than the industry sector in terms of electricity consumption, with 9600TWh consumed in the buildings sector, compared to 7400TWh in the industry sector in 2010.

Electricity consumption in buildings varied significantly across regions in the past 20 years (Fig. 2.4). Globally, it increased on average by 3.7%. North America and Europe grew 2.5%, while growth topped 14% in China, 9% in India, and 7% in the Middle East and Asia. The trend shall continue in the coming 20 years, with an annualized growth forecast of 2.5% (CPS scenario). Other scenarios present similar evolutions. Consumption in North America and Europe is expected to grow by about 1.2% on average. It should slow down in China, although growth should be above 5%. The highest growth would come from India, with above 6% of growth per year. The rest of the world would see growth levels in the range of 3–4%, with the exception of Eurasia, which growth is expected to remain similar to that of OECD

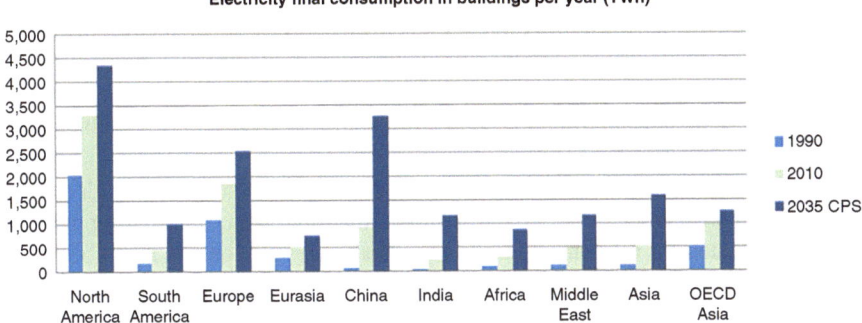

Fig. 2.4 Electricity final consumption in buildings (© OECD/IEA, WEO 2012, 2014)

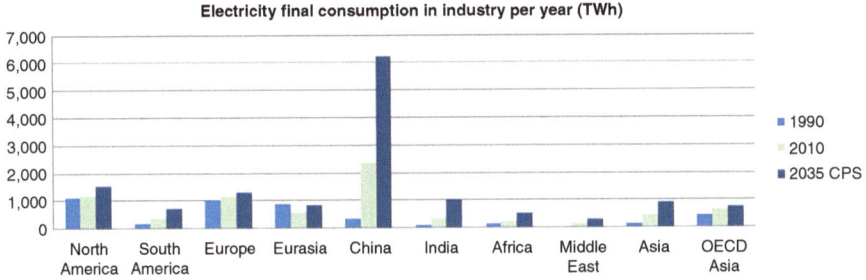

Fig. 2.5 Electricity final consumption in industry (© OECD/IEA, WEO 2012, 2014)

countries, at 1.9%. If the share of OECD countries of the total electricity consumption in buildings was around 80% in 1990, it should only be 47% by 2035, as new economies emerge and increase their living standards. 75% of the electricity consumption growth in buildings shall indeed come from non-OECD countries in the coming 20 years.

The electricity consumption increase in the industry sector in the past 20 years was mainly pulled by non-OECD countries, in particular China (Fig. 2.5). China alone represented 67% of the total growth of electricity consumption in the world, with over 10% of annualized growth against a world average of 2.6%. Other regions also posted high growth. They include the Middle East (6.7%), India (5.8%) and Asia (5.9%), albeit from much smaller baselines. Their contribution to the overall growth has thus remained limited. The growth of electricity consumption in the industry sector has remained limited in OECD countries, at about 0.5% or less in most countries. The CPS scenario confirms this trend in the coming 20 years, with a worldwide growth of 2.8%, similar to the previous period. Other scenarios confirm this trend. The contribution of non-OECD countries to the consumption growth in the sector shall top 90%. China alone would represent around half of the worldwide growth, with an annualized growth of around 4%, though lower than in the past decades. The sustained growth in other regions such as India (5.3%) would also contribute to the worldwide growth of consumption. OECD countries would slowly increase their electricity consumption, by less than 1% per year. While the growth of electricity consumption was slightly more balanced in buildings, the industry sector essentially develops in new economies. With the industry sector accounting for than 59% of electricity consumption in 1990, the share of OECD countries shall drop to less than 26% by 2035.

Finally, the transportation sector represents around 2% of the total electricity consumption worldwide (Fig. 2.6). It is thus a marginal area of activity, predominantly dominated by oil consumption. This sector grew by only 1.8% in the past 20 years from a very low base. Most of the consumption was in Europe and Eurasia, where there has been traditionally some level of electric transportation. CPS projections estimate a growth of this market of 3.3% in the coming 20 years. The 450 scenario estimates growth to be slightly higher, around 5%. In the more aggressive Energy [R]evolution scenario, growth could top 11%, with electricity consumption of the sector to multiply by ten within the coming decades. Overall consumption in the transportation sector would still be much lower (around five times) than in other sectors. In the Energy [R]evolution scenario, this growth would mainly happen in North America, Europe and Eurasia, and finally China, which is expected to post an annualized growth of nearly 8% in the coming 20 years.

In summary, the forecasted annualized growth of electricity consumption in buildings should go from the average 3.7% in the past 20 years to 1.7–2.5% in the coming two decades. This expected slowdown will happen everywhere, notably in new economies. The trend shall be slightly upwards in the industry sector, with growth going from 2.6% in the past 20 years to 2.8% on average in the coming 20 years. The growth is expected to be halved in the Middle East, in China and in Asia, and slightly reduce in India. Obviously the current global economic situation urges caution in terms of forecasts. Forecasts have often proven to be inaccurate, as small variations in key parameters of the model can yield significant variations over time. Currently, the fundamentals of growth remain in China, India, the Middle East and Asia, with much lower consumption per capita than in other regions of the world, and a strong potential for economic and social development.

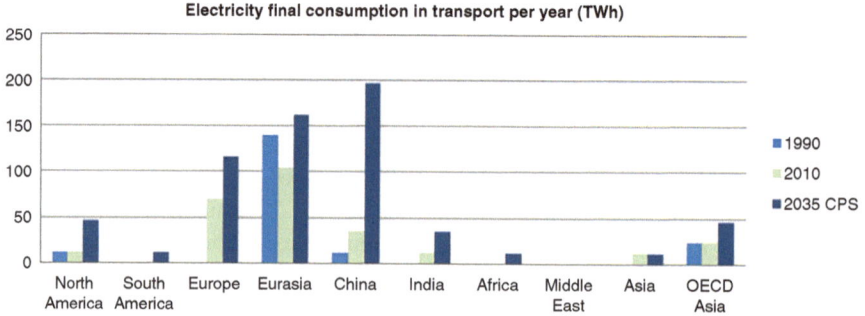

Fig. 2.6 Electricity final consumption in transport (© OECD/IEA, WEO 2012, 2014)

2.2 A Model to Understand the Electricity Consumption Evolution

2.2.1 Four Factors that Impact Overall Electricity Consumption Forecast

As explained in the previous chapter, electricity consumption is expected to increase significantly by over 85% in the coming 20 years (and 53% in the 450 scenario), as a continuation of a trend which could already be seen in the past 20 years. The current forecasts of electricity consumption (notably from the International Energy Agency (2012, 2014)) seem conservative in light of the evolution witnessed in the previous decades, considering the strong short-term potential for development of new economies, such as China, India, the Middle East and Asia, as well as looking at the significant development of the digital economy and its implications. Forecasts of electricity consumption include a variety of factors which, depending on their evolution, impact more or less strongly the overall forecast. This chapter focuses on four of them and provides insight on their relative importance to the overall evolution of electricity consumption. The world population is expanding fast and this is driving the baseline consumption up. Energy efficiency related measures and the renovation of old buildings or the construction of new industrial plants with updated processes lead to a decrease of the energy intensity of buildings and industrial plants, which therefore lead to a decrease of the energy (and thus electricity) consumption. Fuel switching strategies are a key parameter for understanding electricity consumption. Various sources of energy can be used for a given application (heating for instance), among them electricity. The evolution of the share of electricity in the overall energy mix has therefore a strong impact on electricity consumption. Finally, global economic development and in particular the rise of living standards, tied as well to global urbanization, notably in new economies, and the development of the digital economy, has a strong impact on the amount of electricity consumed per individual and therefore total electricity consumption (Fig. 2.7).

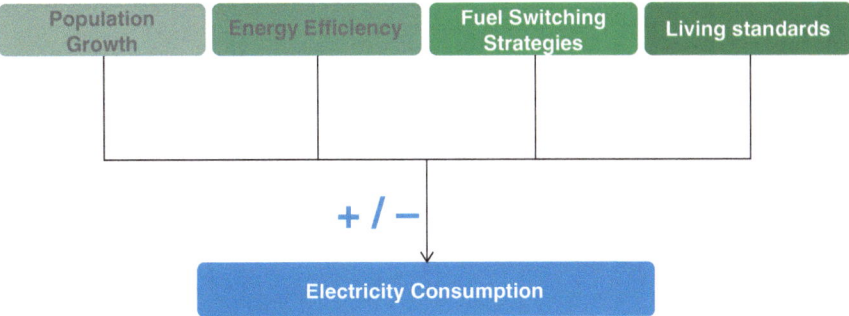

Fig. 2.7 A model for electricity consumption forecasting

2.2.2   *Population Growth*

The world population considerably increased in the past century. The various improvements of sanitary conditions, health services and medicine have led to a demographic transition which happened first in Europe and North America before it disseminated to other regions. This development resulted in a drop in mortality, in particular infant mortality, which led to a massive increase in the population. As economic development yielded better living conditions and higher education levels, the population progressively transitioned to a model with fewer children per family. The demographic transition then ended, and a new one started in which families with a large number of children and a high mortality rate moved to families with fewer children and a low mortality rate. In the meantime, the world population expanded dramatically.

Since the second half of the twentieth century, as the demographic transition spread to most regions of the world, population expansion accelerated. The world population stands at 7 billion people today, compared to only 2.5 billion people in 1950 (Planetoscope 2016; Geohive 2016) (Fig. 2.8a).

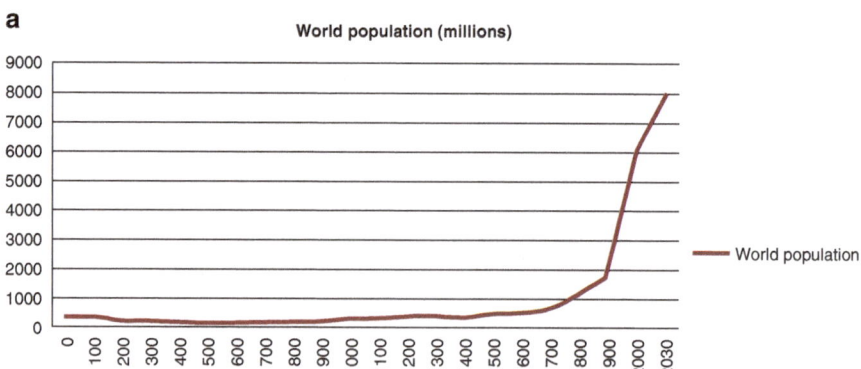

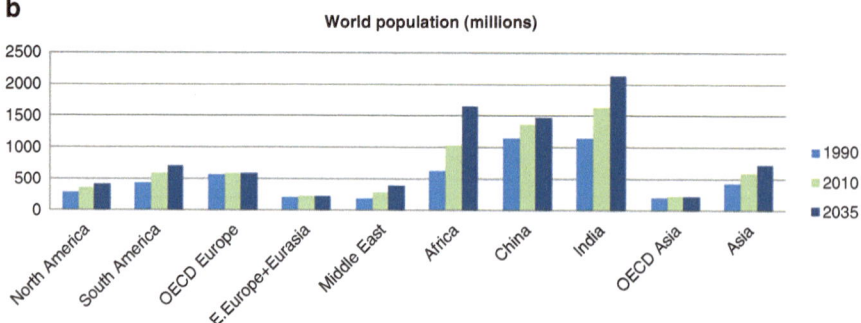

Fig. 2.8 (**a**) World population (Planetoscope 2016; Geohive 2016). (**b**) World population (Our World in Data 2016; Planetoscope 2016)

The world population should increase to 8.6 billion people by 2035 (Our World in Data 2016), and to between 9 and 10 billion people by the middle of the century (Planetoscope 2016).

Most of the world population increase will come from new economies, in particular Africa (almost 2% per year) and India (1.1% per year) (Fig. 2.8b). Other regions such as Asia and the Middle East are also expected to increase by around 1% per year. In China, the population would stabilize, with growth expected to reach a low 0.3% per year. By 2035, the world population would have increased by over 1.5 billion people.

Obviously, a larger world population will impact energy consumption in general and electricity consumption in particular. Assuming no change in the energy mix or in electricity consumption per capita, the impact of the population increase on the electricity consumption can be evaluated by multiplying the baseline of consumption today with the increase in population (Fig. 2.9).

The electricity consumption per capita in North America is about 13MWh/year, almost four times the world average of 2.6MWh/year, so any increase in the population in North America has a massive impact on world consumption. While the population increase in North America will represent only 4% of the world population increase by 2035, it will however represent 38% of the electricity consumption increase. In comparison, Africa represents 37% of the world population increase, but only 14% of the world electricity consumption increase. India represents 30% of the world population increase, and only 8% of the electricity consumption increase.

The increase in the world population is forecasted however to represent no more than 15% of the total electricity consumption growth worldwide. This ratio varies depending on the region. It is very high in the Middle East and Africa (29% and 48%, respectively) and suggests that there is no radical shift in the two regions such as significant fuel switching strategies. Typically, these results indicate that the economic transition in progress is not expected to bring a significant surge of

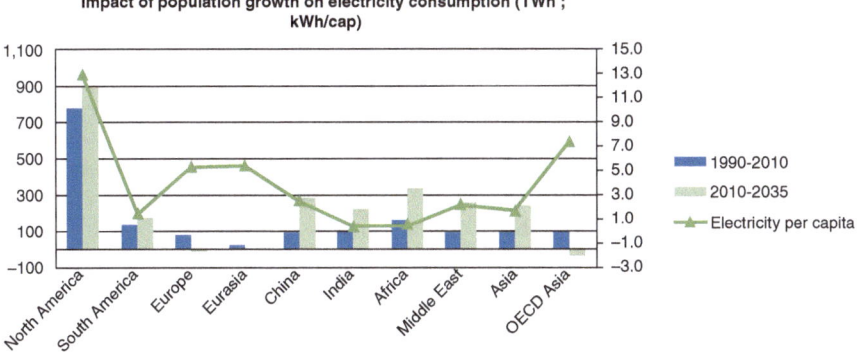

Fig. 2.9 Impact of population growth on electricity consumption (© OECD/IEA, WEO 2012, 2014; Our World in Data 2016; Planetoscope 2016)

electricity consumption up until 2035. On the contrary, forecasts in India and China show a strong increase in electricity consumption in the coming 20 years. The fact that the growth of the population represents a small percentage of the overall growth (13% for India, 4% for China) suggests that these countries are forecasted to either increase significantly their living standards or to change their energy mix towards more electricity (e.g., biomass to electricity), or both. It is also interesting to note that the population increase represents 61% of the electricity consumption growth in North America, a sign of the dependency of the consumption in the region on the population evolution. With a relative low growth of electricity consumption in the region (between 0.6% and 1.4% per year, depending on the scenarios), the evolution of the population represents one of the defining parameters of growth.

2.2.3   Energy Efficiency

A second element that can impact massively the consumption of electricity is the actual energy efficiency measures in the various sectors of activity. This will greatly differ from one sector to another as well as from one region to another, depending on the measures implemented and the current energy efficiency.

The various measures implemented in the past 20 years have led the world to significantly reduce its energy intensity. On average, the world's energy intensity dropped by 0.8% per year between 1990 and 2010 (© OECD/IEA, WEO 2012). Some regions and countries reduced significantly their energy intensity, such as North America and China, while others maintained or even increased their intensity, like in the Middle East for instance. Global energy intensity optimization has actually significantly dropped in the last years of the past decade because of the quick development of new economies and their growing share of the world's energy consumption. Various forecasts all expect energy intensity optimization to accelerate again in the coming decades. In the International Energy Agency's CPS scenario, energy intensity is expected to drop 1.49% per year on average in the coming 25 years. One needs however to be careful here as energy intensity does not depend on energy efficiency measures alone—it also depends on structural and behavioral changes. The Energy Information Administration (2010) generally considers that 25% of the "locked in" reduction corresponds to current policies around energy efficiency. The International Energy Agency (2015) has also recorded the impact of energy efficiency on demand in the past 20 years. According to our calculation, this impact represents as well about 25% of the actual drop in energy intensity. Overall, we consider further that 25% of the "locked in" energy intensity reduction in the CPS scenario corresponds to current energy efficiency measures deployed. This approximation will help us provide an overall perspective of energy efficiency savings. Energy efficiency measures have different impacts on different sources of energy. We have considered here that electricity savings would however be proportionate to overall energy savings. For the 1990–2010 period, the calculation yields savings of about 1300TWh, consistent with the evaluations from the Energy Efficiency Market

Report (© OECD/IEA, Energy Efficiency 2015). It also yields, for the 2010–2035 period in the CPS scenario, electricity savings already "locked in" of around 3300TWh. Additional energy efficiency measures could however be deployed. The NPS scenario estimates these to reach up to 1500TWh (© OECD/IEA, WEO 2012, 2014), following the same principles described above, leading to a reduction of 4800TWh by 2035, or about 15% of the projected electricity consumption by then.

2.2.3.1 Buildings

The buildings sector represents vast potential for energy efficiency. First, the proper insulation of buildings can yield massive energy consumption savings. According to the International Energy Agency (© OECD/IEA, Buildings 2013), up to 60% of heat consumption could be saved through the installation of best-in-class insulation. The thermal insulation of buildings through the use of windows is generally considered to have a favorable impact of up to 10% over space heating (© OECD/IEA, Efficient Buildings 2013). Appliances and lighting, which represent up to 25% of the energy consumption in a building (© OECD/IEA, Buildings 2013), can also be optimized. Sealing improvements on boilers, refrigerators and washing machines have already resulted in significant improvements over the last decades. For instance, the electrical consumption of refrigerators has dropped by 30% in the last 10 years. LEDs (light emitting diodes) are progressively replacing traditional lamps for lighting, yielding savings of more than 50% (© OECD/IEA, Energy Efficiency 2013; ADEME/Energy Efficiency 2013). Beyond these improvements, energy management technologies strongly contribute to energy savings. "Active controls", designed to operate in real time, optimize energy consumption as close as possible to end users' requirements. According to Schneider Electric (2014), they can reduce from 10% to 60% of the energy consumption of a building. According to the ADEME/ Energy Efficiency (2013), these savings could reach up to 50% for appliances, and beyond that for motors.

Overall, energy efficiency projections based on data from the International Energy Agency (2012) point to around 1900TWh of savings already "locked in" the CPS scenario, and a possible additional 450TWh of savings in the NPS scenario. None of these two scenarios however assumes full realization of the energy efficiency potential. The International Energy Agency (2012) has estimated the economic viable potential for energy efficiency. Following the same assumptions described above, the full potential could reach around 2500TWh of additional reduction on top of the CPS scenario, or 4400TWh of savings overall, around 25% of the forecasted consumption in the sector in 2035. Now, this value may seem conservative when looking at the actual energy consumption of new buildings. The ADEME/Energy Efficiency (2013) has estimated that while existing buildings consume around 400kWh/m^2, renovated buildings can consume as low as 80kWh/m^2, and new buildings could end up being net energy producers. Considering the potentials described above, we will assume here a full potential of 30% of savings on

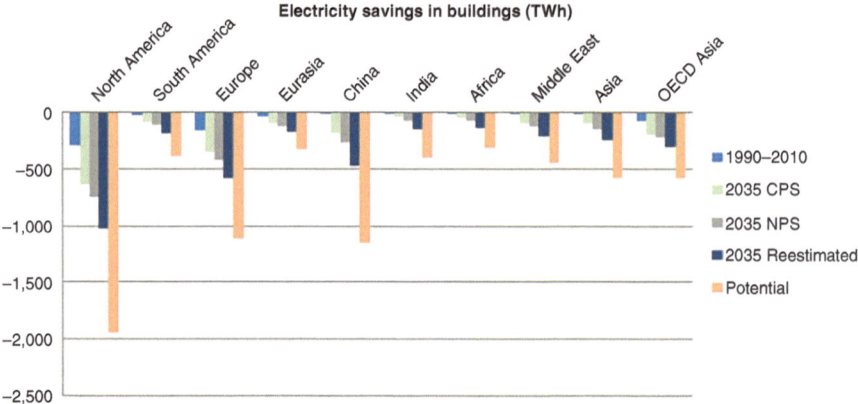

Fig. 2.10 Electricity savings in buildings (ADEME/Energy Efficiency 2013; © OECD/IEA, WEO 2012, 2014)

electricity going forward. One main issue however is the actual low renovation rate of existing buildings. By 2050, 75% of existing buildings are still expected to be standing. It is thus clear that ambitious renovation plans and new building constructions could target reducing the energy footprint of buildings by a significant factor, as a minimum in OECD countries, which have the highest energy density. Assuming 30% potential reduction in all buildings around the world, the calculation yields a potential as high as 7200TWh. This estimate of the potential will be the one used further on. Again, the energy efficiency savings estimate from the International Energy Agency seems conservative, especially in light of the evaluated potential. A re-estimated forecast was thus plotted here (Fig. 2.10) with the assumption that 30% of the potential would be realized in various countries by 2035. The re-estimated forecast yields savings of around 3500TWh (against 2300TWh in the NPS scenario and 1900TWh in the CPS scenario). North America stands out with 30% of the worldwide savings, followed by Europe with 17%.

2.2.3.2   Industry

The industry sector also represents a vast potential for energy savings. Experts agree that up to 30% of the energy could be saved in petrochemical plants, up to 20% in steel plants, around 25% in cement plants and 13% in aluminum plants (© OECD/IEA, Technology Industry 2009) if best-in-class technologies were properly deployed. These technologies include optimized process control, the use of combined heat and power to maximize both reuse of generated heat and the yield of process operations, the reuse of scrap and waste and, in some cases, the redefinition of the processes themselves. Beyond primary industries, electrical motor systems also represent an important source of optimization. They indeed represent close to 70% of the total electricity consumption in the industry (© OECD/IEA, WEO 2012, 2014). Up to 20% of electricity could be saved in efficient electric motor systems

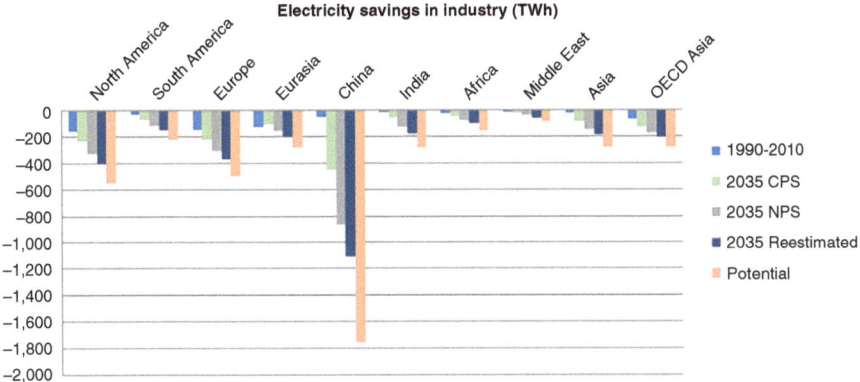

Fig. 2.11 Electricity savings in industry (© OECD/IEA, Technology industry 2009; © OECD/IEA, WEO 2012, 2014)

compared to traditional systems, leading to electric motor systems with efficiency of up to 63% (© OECD/IEA, Motors 2011). Overall, if we use the same data source and assumptions for electricity savings for buildings, the total potential of electricity that could be saved in the industry sector amounts to up to 3600TWh in 2035. This represents around 25% of electricity consumption by the sector in 2035. Assuming that about 20% of savings could be achieved by motors beyond other measures of efficiency, we reach a slightly higher full potential of around 4400TWh of electricity savings. This figure will be used further on.

The International Energy Agency (2012) estimates that, by 2035, the equivalent of 1400TWh of electricity consumption could be saved in the industry sector (CPS scenario), and up to 2400TWh in the more energy efficient NPS scenario. Existing plants are expected to be progressively renovated in the coming decades, and a number of them decommissioned. Also, new plants are expected to benefit from the latest standards in terms of energy efficiency. Assuming 50% of the potential would thus be realized worldwide in all industries by 2035, a concurrent forecast yields potential savings of 2900TWh in the industry sector (Fig. 2.11). China obviously stands out with around 40% of the worldwide savings.

2.2.3.3   Summary

The transportation sector has also a massive potential for energy savings. Since electrical consumption in the sector is negligible, it is considered however that electricity savings are close to none. Estimates based on data from the International Energy Agency yield nevertheless around 100TWh, an insignificant number compared to the buildings and industry sectors. In summary, the re-estimated potential of electricity savings from energy efficiency measures amounts to 11,700TWh, which represents more than one third of the electricity consumption forecast for 2035. The forecast by the International Energy Agency (2012) is conservative, with

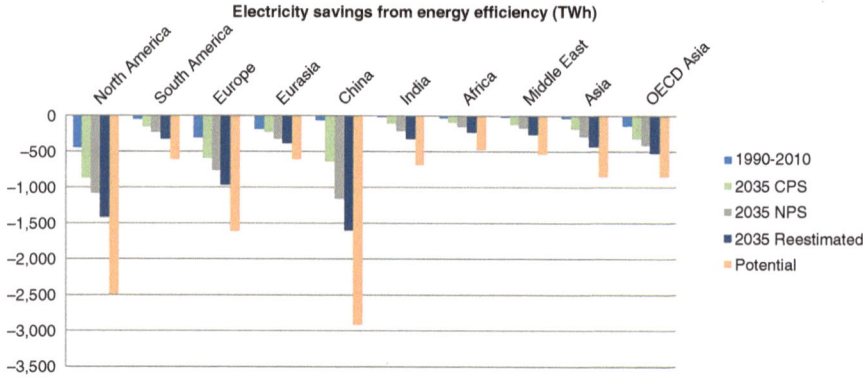

Fig. 2.12 Electricity savings from energy efficiency (ADEME/Energy Efficiency 2013; © OECD/ IEA, Technology Industry 2009; © OECD/IEA, WEO 2012, 2014)

3300TWh–4800TWh of savings by 2035, depending on the scenarios. Our concurrent forecast estimates that up to 6400TWh could be saved (Fig. 2.12).

2.2.4 Fuel Switching Strategies

Fuel switching strategies can also influence greatly electricity consumption. They refer to the substitution of fossil fuels (oil, natural gas, coal) or traditional biomass by electricity as a source of energy. For buildings, countries relying on oil, coal or natural gas for heating can decide to move to electricity, in particular renewable electricity. In new economies, traditional biomass used for heating, cooking or lighting can also be substituted by electricity (from power networks' natural expansion in rural areas or from microgrids), such as in India or Africa. Today 19% of the world population does not have access to electricity (© OECD/IEA, WEO 2012), particularly in India, South East Asia and Africa. By 2030, the International Energy Agency (2012) plans to reduce this ratio to 12%, which corresponds, taking into account the concurrent increase in the world population, to two additional billion people. In the industry sector, fossil fuels used in traditional heating processes can also be substituted by either biomass or electricity. One example is the substitution of traditional blast furnaces in steel manufacturing by electric arc furnaces. The International Energy Agency (2012) has presented the evolution of all sources of primary energy used by each sector between 1990 and 2010, as well as estimates until 2035. The evolutions of energy mix can thus be evaluated. The share of electricity is growing everywhere during the period 2010–2035, to the detriment of fossil fuels and to traditional biomass in buildings in new economies (Fig. 2.13).

It is clear that electricity is taking a growing share of the energy supply in buildings and in the industry sector. Traditional energy supply types used in buildings, such as fossil fuels (with the exception of natural gas) and renewable

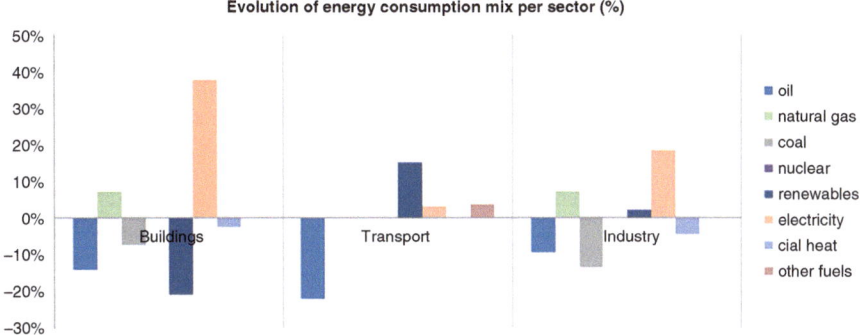

Fig. 2.13 Evolution of energy consumption mix in sectors (© OECD/IEA, WEO 2012)

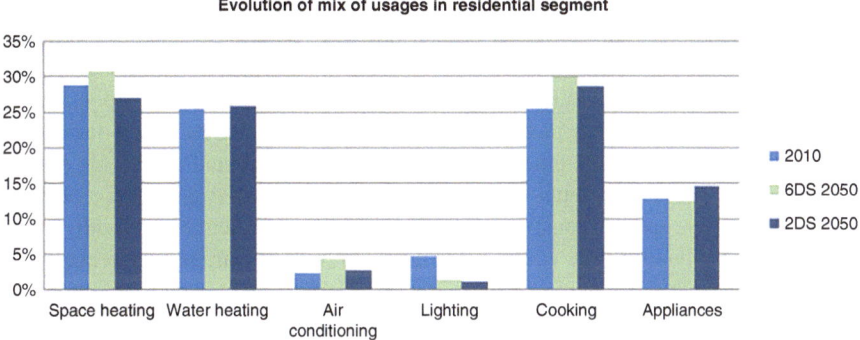

Fig. 2.14 Evolution of the mix of usages in the residential segment (© OECD/IEA, Buildings 2013)

(traditional biomass), are on the decline. In the industry sector, electricity mainly replaces fossil fuels (again with the exception of natural gas) as well as commercial heat. In the transportation sector, oil is losing share to a mix of solutions, particularly renewables, which here corresponds to biofuels.

In principle, these changes in mix cannot only be attributed to fuel switching strategies, as usages also evolve (© OECD/IEA, Buildings 2013). According to the International Energy Agency (2013), the share of air conditioning, lighting and appliances in the residential segment (with 100% electric supply) do not vary significantly whatever the scenario (we will come back to this assumption when looking at the impact of living standards). Other usages such as heating and cooking, which rely on electricity alone thus keep the same shares (Fig. 2.14).

From this it can be concluded that most of the evolutions of the electricity mix are actually related to substitution strategies for heating and cooking usages, between traditional biomass (renewable), fossil fuels, and electricity supply. The tertiary segment (which includes all services businesses and basically all commercial and industrial buildings) offers a similar perspective, with a share of electric appliances which varies around 60%, depending on the scenario. The industry and the transportation sectors' energy consumption being mainly associated with processes, all variations of mix can reasonably be associated with fuel switching strategies in powering processes.

The variations of the energy mix in buildings show that around 5500TWh of electricity could be additionally consumed by 2035 as the result of switching from fossil fuels or traditional biomass to electricity. This is slightly higher than in the past 20 years (3400TWh). This would represent around 30% of the overall electricity consumption in buildings by 2035, and over half of the increase in electricity consumption since 2010. China and India would be the primary countries where this change would occur, as the result of the conversion from traditional biomass used for heating and lighting to electricity. The move to electricity supply would also be faster in China than in India. OECD countries would also experience slight changes of mix, mainly due to the move from old heating systems to electricity supply, as well as the development of distributed renewable energies as a replacement for natural gas. These forecasts however do not reflect the potential of fuel switching strategies in the various regions. Indeed, there is still a lot of potential to convert traditional biomass into electricity as well as substitute fossil fuels with electricity, notably with the emergence of affordable distributed renewable power. Theoretically, the full amount of traditional biomass and fossil fuels could be substituted. This is obviously a highly theoretical calculation, with about 27,700TWh of energy that could be substituted with electricity. However, this does not represent the volume of electricity that would be generated additionally. Indeed, electricity and other solutions have different efficiencies when it comes to "useful" energy, or the actual energy consumed for a given mechanical or thermal energy service. We focus here on electric heating, which is the dominant usage to be substituted. Electric heating systems are said to have efficiency ratios of 100%, as opposed to natural gas heating, which range around 85% (DOE 2017). The comparison between those two sources of energy is the one traditionally used, and it applies well to heating-intensive systems, industrial applications, and buildings in very cold climates. In countries with milder climate (the dominant form of building applications around the world), this comparison fails to illustrate how electric technologies enable smarter and more efficient heating in buildings. Indeed, electric heating can be regulated better. Various studies have shown that electric heating can be up to three times more efficient than natural gas heating for a similar level of comfort (Observatoire de l'energie 2004; CEREN 2009). Additionally, heat pumps can yield up to three or four units of heat for one unit of energy effectively consumed. Heat pumps present a significant potential as they also enable the cooling of the atmosphere when the weather is hot. With the development of HVAC systems, reversible HVAC systems offer an interesting alternative to having two separate systems for heating and cooling, with significant efficiency ratios. The ratio of efficiency considered here for electric heating is 200%, or slightly more than twice the efficiency of natural gas heating. Another way of looking at it is, for 100% of electric heating efficiency, the efficiency of natural gas heating is divided by two, down to around 42%. This is the ratio that will be considered further on. In total, up to 17,100 TWh of additional electricity could be required if biomass and fossil fuels were replaced by electricity. This would represent 1.5 times the total electricity consumption in the buildings sector in 2035. A large potential thus remains for an increase in electricity consumption in this sector. A concurrent forecast could thus be added to the one from the International Energy Agency (2012). This forecast estimates that 70% of coal and oil, 30% of traditional biomass, and 5% of natural gas and commercial heat could be substituted by 2035. This would lead to a higher forecast, around 8700 TWh of additional electricity consumption overall (Fig. 2.15).

In the industry sector, the variations of mix show that 3000 TWh could be additionally consumed as electricity by 2035 as the result of fuel switching from fossil fuels. This is slightly more than what was experienced in the last 20 years (2200 TWh). China alone would represent two thirds of the fuel switching, and India 11%, following a massive modernization of the industry footprint in the region. This increase in electricity consumption would correspond to 21% of the total electricity consumption in the sector in 2035 and close to half of the overall increase since 2010. The remaining potential is however related to the current consumption of fossil fuels in the industry sector. It is possible to retrieve data for the actual consumption for each fuel type in each industry sector from the International Energy Agency (2014). This helps in the understanding of the actual industry mix of each region, which varies greatly from one another (Fig. 2.16).

The Energy Transition Commission evaluated the potential of electrification in industry sectors in 2017. It notably showed that electro-thermal technologies have the potential to enable significant process electrification in various industries:

– Paper industry: 85% on drying processes, which account for 60% of total energy (Springer 2012)
– Aluminum industry: 25% on bauxite reduction, which represents 21% of total energy (TUC 2012)

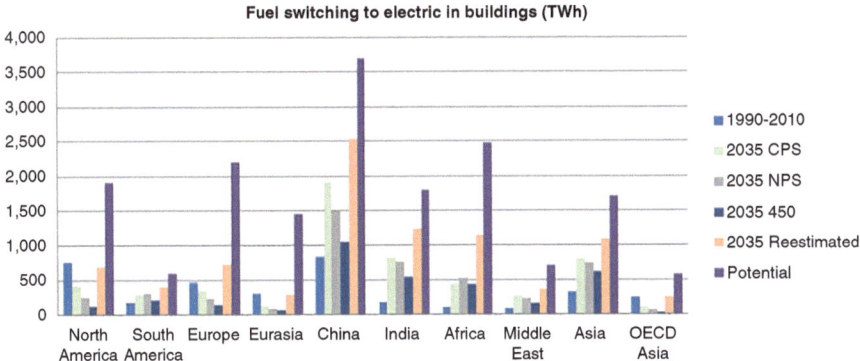

Fig. 2.15 Fuel switching to electricity in buildings (© OECD/IEA, Buildings 2013; © OECD/IEA, WEO 2012, 2014)

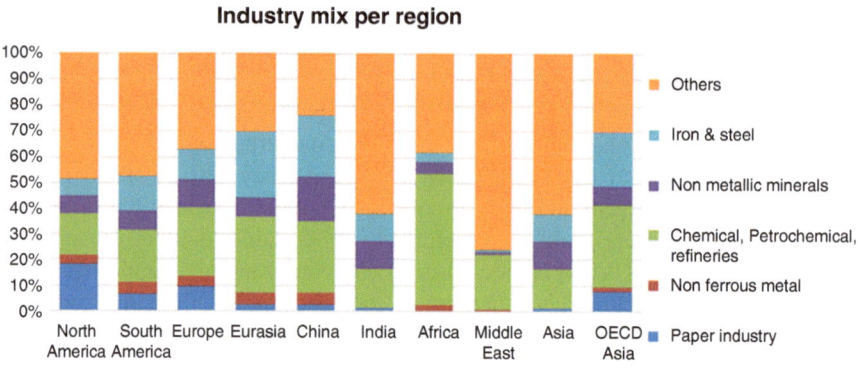

Fig. 2.16 Industry mix by region (© OECD/IEA, Sankey 2014)

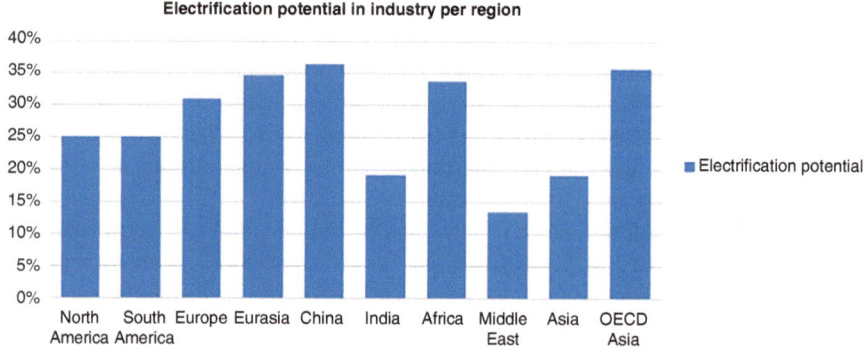

Fig. 2.17 Electrification potential in industry (© OECD/IEA, Sankey 2014; ETC 2017)

– Petrochemical industry: 65% on cracking process, which, together with boilers, represents around 90% of energy (DOE/Manufacturing Energy 2013).
– Cement industry: 80% on clinker process, which represents 53% of total energy (JCA 2018)
– Steel industry: 25% on recycled steel, 70% on primary steel production with direct ore reduction, and an evolution of mix towards more recycled steel (50% potential assumed).

Overall, the industry sector has a global potential of 29% from fuel switching to electricity, with regional variations between 13% (Middle East) and 36% (China) (Fig. 2.17).

Assuming a 85% efficiency ratio between electric and traditional heating as explained earlier, this leads to a potential of 8100TWh of additional electricity consumption in the industry sector, or 60% of the forecasted consumption by 2035. This shows that a large potential exists for electricity consumption to increase in the sector. A concurrent forecast could thus be provided that considers 40% substitution of oil and renewables, 70% of coal, and 5% of natural gas and

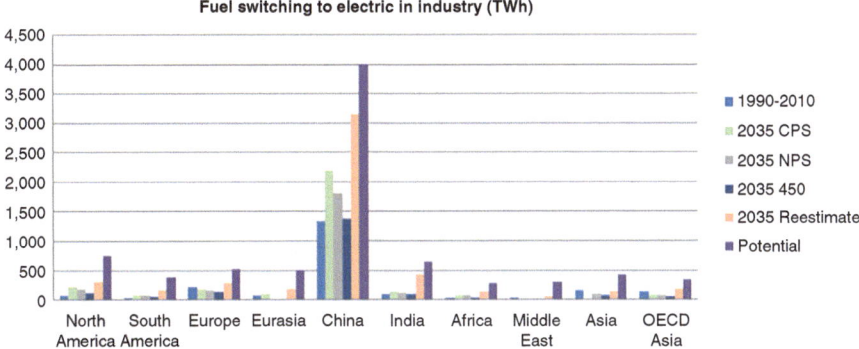

Fig. 2.18   Fuel switching to electric in industry (© OECD/IEA, WEO 2012, 2014)

commercial heat. This concurrent evaluation would lead to 5000TWh of additional electricity consumption in the sector (Fig. 2.18).

Finally, the forecasts in the transportation sector show that up to 220TWh of additional electricity could be consumed from switching from fossil fuels to electricity in the form of electric vehicles. The share of OECD countries would represent 40% of this additional demand, while China would represent close to the remaining 60% of the increase. By 2035, this switch from fossil fuels to electricity in transportation would represent half of the increase in consumption in the sector, and up to 30% of total electricity consumption in the sector. Now, this would still represent a minor share of the total energy consumption of the sector. As mentioned earlier, the scenarios from the International Energy Agency (2012) are very conservative. In comparison, the Energy [R]evolution scenario estimates consumption could go as high as 3800TWh, with similar regional disparities. The actual potential is actually quite larger than what forecasts estimate. A calculation, detailed in the chapter dedicated to electrical vehicles, yields a full potential of around 5000TWh of additional electricity consumption. This calculation considers that theoretically all short-distance transportation by car could be migrating towards electric vehicles. Now, the replacement rate of cars is very low, around 7% (© OECD/IEA, Transport 2009). A concurrent forecast could thus assume that up to 50% of the full potential would be realized by 2035 in OECD countries and China, and 30% in other regions, leading to around 2000TWh additional consumption from electric vehicles (Fig. 2.19).

In summary, fuel switching strategies are forecasted to generate an additional demand for electricity of nearly 8700TWh by 2035 in the CPS scenario, out of a total increase of 15,700TWh over the period, or 55% of the increase. A concurrent forecast could be built around 15,700TWh of additional consumption from fuel switching strategies, mainly from buildings and industry segments. Non-OECD countries would be the primary regions concerned, with notably an accelerated substitution of biomass by electricity (in buildings) and stronger development in the industry sector (Fig. 2.20).

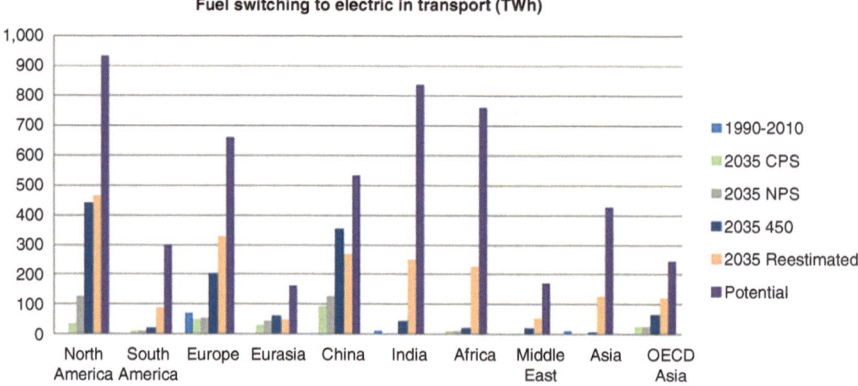

Fig. 2.19 Fuel switching to electricity in transport (Greenpeace 2015; © OECD/IEA, WEO 2012, 2014)

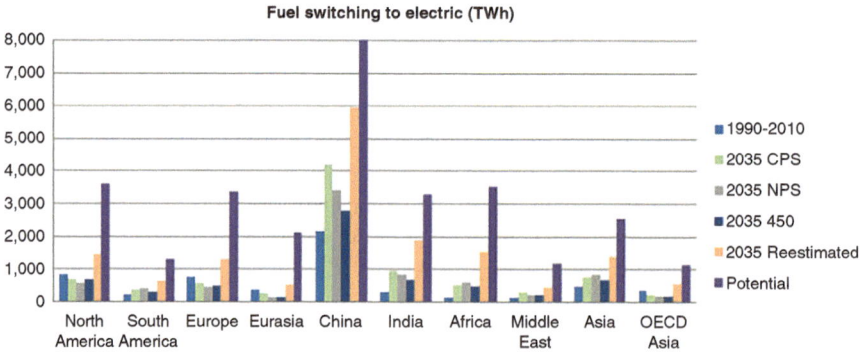

Fig. 2.20 Fuel switching to electricity (Greenpeace 2015; © OECD/IEA, Buildings 2013; © OECD/IEA, WEO 2012, 2014)

2.2.5 Living Standards

The final element that helps explain the electricity consumption evolution is the actual evolution of living standards due to economic development. This global contribution can be estimated by deducting the contributions described above from the actual CPS scenario forecasts. Overall, the contribution of living standards to electricity consumption would increase in the coming period (7800TWh) compared to the previous period (2000TWh). Non-OECD countries would represent two thirds of the growth, and China alone 30% (Fig. 2.21). In absolute value terms, the rise of living standards would contribute to more than half of the growth of electricity consumption between 2010 and 2035. The impact would be significant in all mature economies, as well as in China. In other regions, the impact is expected to generate a

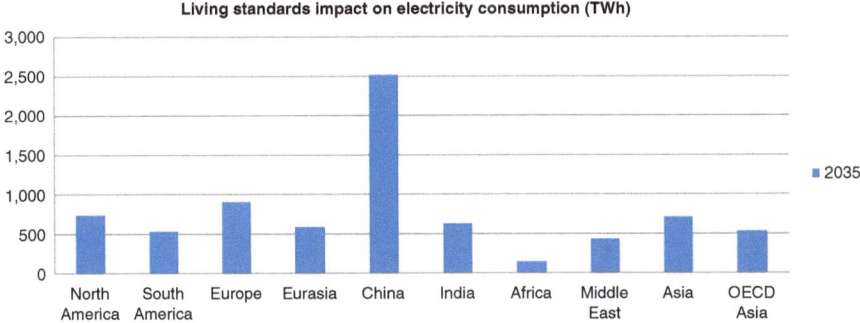

Fig. 2.21 Living standards' impact on electricity consumption (Author's own calculation based on data from International Energy Agency)

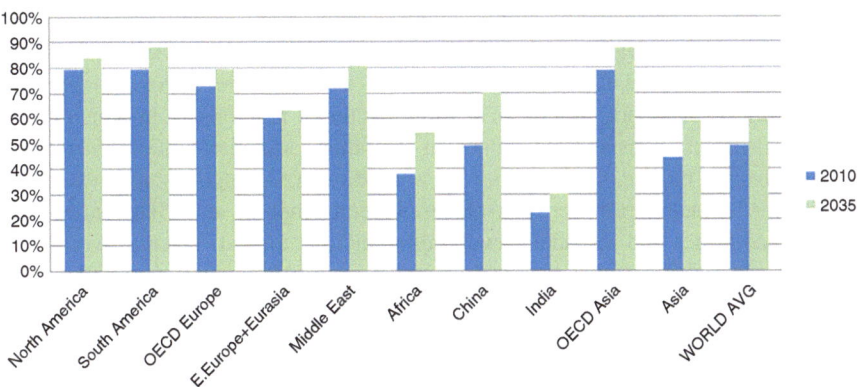

Fig. 2.22 Evolution of world population urbanization (UN 2014)

lower increase in electricity consumption in absolute value, also a sign on the dependency between electricity consumption and Gross Domestic Product (GDP).

This forecast remains however very approximate as the actual model is very complex. One of the main elements which contribute to the growth is the level of urbanization. This is expected to grow by 10% on average in the coming 20 years (UN 2014) to reach an average of 60% in 2035, compared to 29% in 1950. The increase should be particularly noticeable in non-OECD countries, which should represent 92% of the world's urban population growth (Fig. 2.22). There should be more than 400 million people additional living in cities in China within the next 20 years, and over 500 million in Africa.

A second element to which the growth of electricity consumption can be compared is GDP growth (Fig. 2.23). According to the Energy Information Administration (2014), global GDP is set to more than double in the coming 20 years to reach above 173 trillion USD in 2035, compared to around 70 trillion USD in 2014.

Several studies have attempted to find a statistical relationship between GDP evolution and electricity consumption; most of them followed Granger's causality

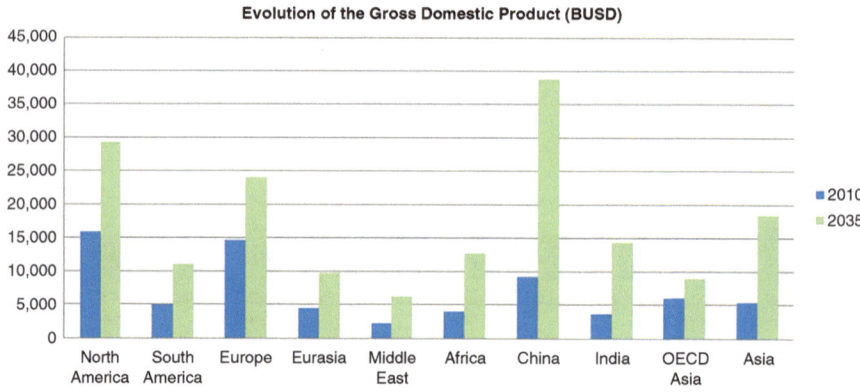

Fig. 2.23 Evolution of Gross Domestic Product (EIA/GDP 2014)

studies (Granger 1969; Klran and Guris 2009; Abaidoo 2010; EIA/Economy and Electricity 2013; Bruns et al. 2013; Hou 2009). All these studies prove a relationship between the two factors. Some show a unidirectional relationship in which either GDP or electricity consumption influences the other parameter, without the reverse being proven to be true. Other studies show a bidirectional relationship between the two factors. The relation between electricity consumption and GDP varies also between countries and the actual evolving economic context. Overall, worldwide electricity annualized growth was 3.2% over the period 1990–2010, for GDP growth of 5.9% and a GDP/capita growth of 4.5%. The forecasts for the 2010–2035 period show an annualized electricity consumption growth of between 1.7% and 2.5% for a GDP growth of 3.7% on average, and a GDP/Capita growth of 2.8%. If we estimate the growth of electricity consumption by projecting a similar trend from the previous 20 years, following the estimates of GDP growth and GDP/Capita growth evolutions for the coming 20 years, we reach in both cases an estimated growth of electricity production of 2% per year, which is close to the actual estimates.

To summarize, urbanization and economic development naturally lead to both an increase and an acceleration of electricity consumption, which represents a significant part of the electricity consumption evolution. We have considered further that this impact would not change depending on the concurrent forecasts of fuel switching strategies and energy efficiency measures; this assumption is challengeable. One significant impact on living standards has also been left unattended in forecasts: the rapid development of the digital economy. Digitization of the economy leads to a surge in electricity consumption. The energy footprint of digitization is made of the electricity consumed by data centers, communication networks, digital appliances and the manufacturing plants that produce these appliances. Overall, it represented in 2013 around 2000TWh, or around 10% of total electricity consumption (Mills 2013). According to the same source, this figure could go up to 6000TWh by 2030. This increase is generally not included in existing energy forecasts. A concurrent forecast for the impact of living standards would then yield 11,800TWh by 2035.

2.3   Summary: The Transition to a New Electric World

Electricity consumption is forecasted to rise by 53–85% within the next 20 years, depending on the scenario. China is expected to represent between 35% and 41% of the total growth of electricity consumption, followed by India (10–18%) and Asia (11–13%). Electricity consumption per capita is expected to increase by 50% in 20 years, similar to the growth witnessed in the past 20 years. India would almost triple its consumption per capita, followed by China and Asia, where it is expected to double. Half of the electricity consumption growth will come from the buildings sector. Within this sector, 75% of the growth will come from non-OECD countries. China alone would represent more than 25% of the total electricity consumption growth. Another 44% will come from the industry sector, with China representing over half of this growth. The transportation sector would account for a negligible increase of 300TWh of consumption by 2035. Overall, electricity consumption would jump from 17,800TWh in 2012 to between 27,000TWh and 33,000TWh in 2035, depending on the scenarios (Fig. 2.24).

Four factors will contribute to this spectacular growth in electricity consumption. First, the world population continues to grow. It should top 8.7 billion people by 2035, compared to roughly 7 billion people in 2010. This increase naturally leads to an increase in electricity consumption. The increase is all the more important in consumption per capita terms. One person in North America typically consumes five times more electricity than one person in China, and close to 40 times more than one person in India! The increase in population forecasted in North America is thus a key contributor to worldwide electricity consumption. With only 4% of the worldwide population growth, North America will represent 38% of the increase in electricity consumption related to the growth of population. In comparison, India, which will represent 30% of the worldwide population growth, will only represent 8% of the electricity consumption growth associated with population increase. Overall, the

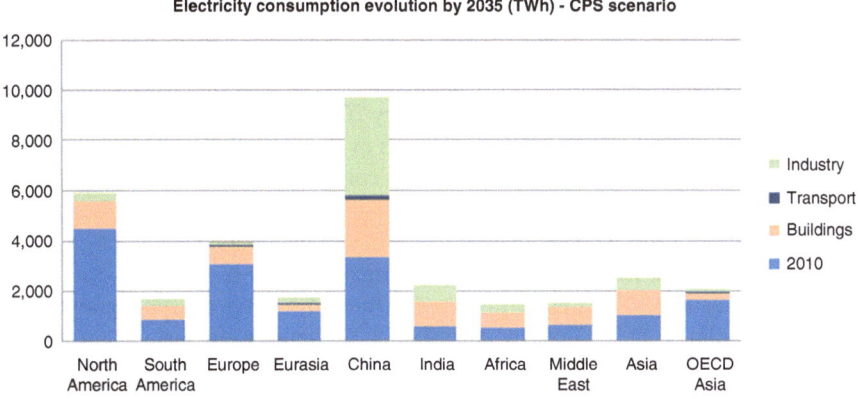

Fig. 2.24   Electricity consumption evolution by 2035 (© OECD/IEA, WEO 2012, 2014)

growth of population could represent an impact of around 2300TWh of electricity additionally consumed, a small share of the overall increase.

Energy efficiency measures are also an important factor that can explain the actual electricity consumption evolution. According to the forecasts based on available data from the International Energy Agency (2012), energy efficiency measures could represent a decrease of 3300TWh in electricity consumption by 2035 in the CPS scenario. Energy efficiency measures in buildings (particularly in North America) would represent more than half of it, while measures in industry (essentially in China) would represent the remaining share. There is however much more to do to unlock the full potential of energy efficiency, in particular in buildings in OECD countries. A concurrent forecast would lead to a 6400TWh or so decrease in electricity consumption by 2035 attributed to energy efficiency measures.

A third and critical factor is the actual change in the energy supply mix. The share of electricity in buildings should indeed dramatically increase to the detriment of oil and renewable energy (traditional biomass), mainly as a result of progressive rural electrification, which would lead to a massive increase in the population having access to electricity (two billion additional people connected to the electrical network) as well as the deployment of affordable solar and wind power. Additionally, the deployment of new (more electricity based) industrial set-ups would also lead to significant evolutions of the mix in the segment. According to the forecasts from the International Energy Agency (2012), fuel switching strategies could represent half of the total increase in electricity consumption by 2035, with an additional 8700TWh consumed. Most of this growth would come from rural electrification in new economies (50%) and fuel switching during the modernization of industrial processes, notably in China (40%). Concurrent forecasts consider that fuel switching strategies could be more aggressive, in particular in buildings in OECD countries and in electric transportation, raising the forecast to 15,700TWh of electricity additionally consumed by 2035.

Finally, data from the International Energy Agency (2012, 2014) suggests that the increase in living standards could impact electricity consumption by around 7800TWh by 2035, with non-OECD countries representing two thirds of this. A concurrent forecast integrating the impact of the digitization of the economy yields 11,800TWh of additional consumption by then.

In the CPS scenario, the increase in population would account for 15% of the total consumption growth by 2035, energy efficiency 20%, while fuel switching strategies would represent 55% of the total consumption evolution and living standards 52%, leading to a total increase of 85% of electricity consumption by 2035 (Figs. 2.25 and 2.26).

As important as this increase may be overall, the remaining potential for growth is all the more impressive. The concurrent estimate yields higher ratios and total electricity consumption (considering similar forecasts for population growth and living standards) of 40,700TWh by 2035 (Fig. 2.27). Energy efficiency has a slightly higher impact (28% of the 2035 consumption evolution) here, particularly in OECD countries, but fuel switching strategies are also further enhanced (in particular in

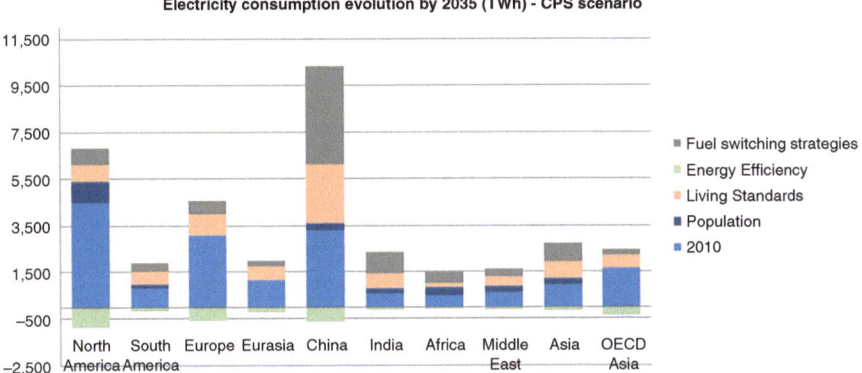

Fig. 2.25 Electricity consumption evolution by 2035—CPS scenario (Source: author's own calculation based on data from IEA)

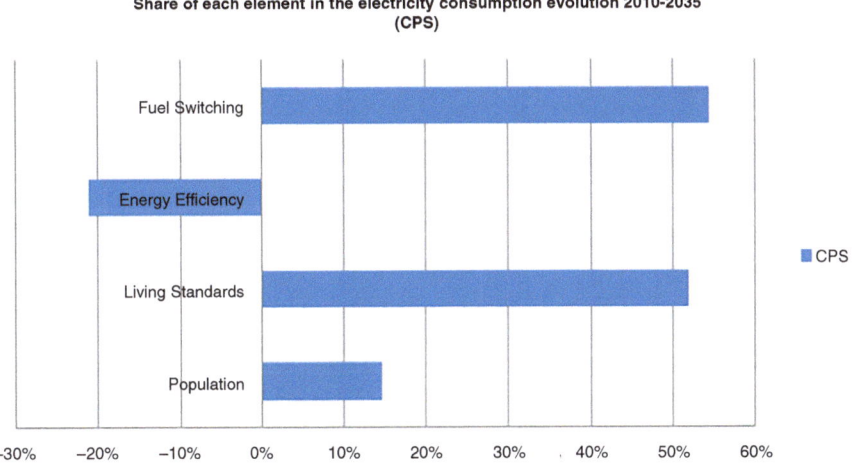

Fig. 2.26 Share of each element in the electricity consumption evolution (Source: author's own calculation based on data from IEA)

non-OECD countries), largely compensating the savings realized (70% of the 2035 consumption increase) (Fig. 2.28).

From a regional standpoint, the growth in electricity consumption would remain limited in OECD countries and the Middle East as a result of energy efficiency deployment, although the limitation would be partially compensated by fuel switching. Current consumption in OECD countries has remained stagnant or even shrunk slightly, a sign of the effectiveness of energy efficiency measures and the lack of aggressive electrification. Time will tell if more of the substitution potential will be realized or not. In other regions of the world, the potential for growth is

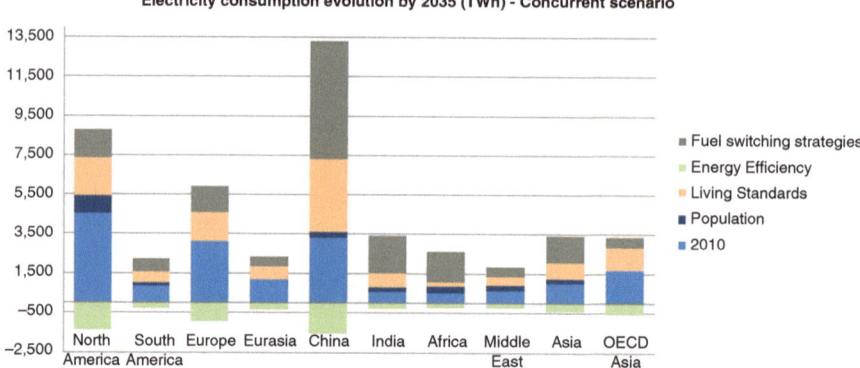

Fig. 2.27 Electricity consumption evolution by 2035—concurrent scenario (Source: author's own calculation)

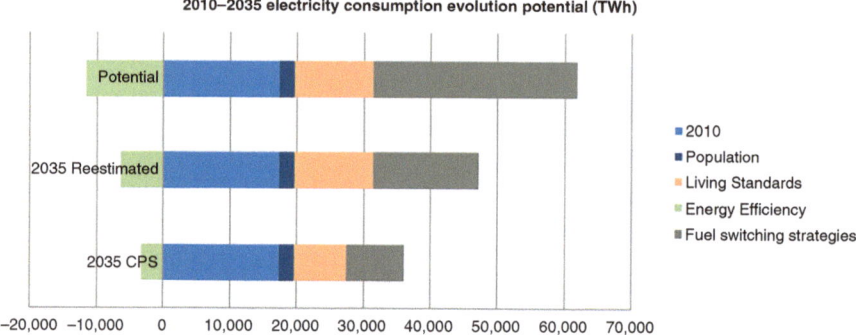

Fig. 2.28 Electricity consumption evolution by 2035 (Source: author's own calculation)

significant, and current forecasts could consequently be revised upwards in the coming years, particularly for India and Africa (Fig. 2.29).

To sum up, general economic development, the digitization of the economy, rural electrification in new economies (Asia, Africa and India), the substitution of fossil fuels with electricity in buildings (in OECD), in industry (notably China), and the faster penetration of electrical transportation in cities are the main factors which need to be monitored to understand the upcoming dynamics of the worldwide electrical consumption evolution. A very significant potential exists for transitioning energy sources to electricity, and electricity will likely gain a bigger share in the worldwide energy mix in the coming decades. While consumption in OECD countries has stagnated in recent years, thanks notably to the impact of energy efficiency measures, the potential for replacing other sources of energy with electricity (particularly in buildings) suggests that consumption could keep increasing in the coming decades. In other countries, the potential is spectacular as population increases, living standards improve, and traditional biomass progressively gives way to electricity. Consumption is thus expected to skyrocket in most of these geographies.

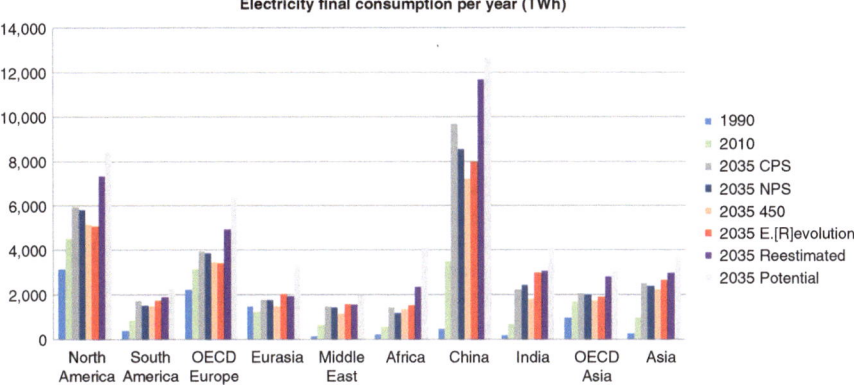

Fig. 2.29 Electricity final consumption per year (Greenpeace 2015; © OECD/IEA, WEO 2012, 2014)

References

© OECD/IEA, Buildings. (2013). http://www.iea.org/publications/freepublications/publication/transition-to-sustainable-buildings.html

© OECD/IEA, Efficient Buildings. (2013). http://www.iea.org/publications/freepublications/publication/TechnologyRoadmapEnergyEfficientBuildingEnvelopes.pdf

© OECD/IEA, Energy Efficiency. (2013). http://www.iea.org/publications/freepublications/publication/energy-efficiency-market-report-2013.html

© OECD/IEA, Energy Efficiency. (2015). *Energy efficiency market report.* http://www.iea.org/publications/freepublications/publication/MediumTermEnergyefficiencyMarketReport2015.pdf

© OECD/IEA, Motors. (2011). https://www.iea.org/publications/freepublications/publication/EE_for_ElectricSystems.pdf

© OECD/IEA, Sankey. (2014). *Sankey diagram.* https://www.iea.org/Sankey/

© OECD/IEA, Technology Industry. (2009). http://www.iea.org/publications/freepublications/publication/industry2009.pdf

© OECD/IEA, Transport. (2009). *Transport, energy and CO2.* https://www.iea.org/publications/freepublications/publication/transport2009.pdf

© OECD/IEA, WEO. (2012). http://www.worldenergyoutlook.org/publications/weo-2012/

© OECD/IEA, WEO. (2014). *In Greenpeace/The energy [R]evolution scenario.* http://www.greenpeace.org/international/Global/international/publications/climate/2015/Energy-Revolution-2015-Full.pdf and https://www.iea.org/publications/freepublications/publication/WEO2014.pdf

Abaidoo, R. (2010). *Economic growth and energy consumption in an emerging economy: Augmented Granger causality approach.* http://www.aabri.com/manuscripts/11843.pdf

ADEME/Energy Efficiency. (2013). *Roadmap for positive-energy and low-carbon buildings and building clusters.* http://www.ademe.fr/sites/default/files/assets/documents/84671_roadmap_for_positive_energy_and_low_carbon_buildings_and_building_clusters.pdf

BP. (2016). http://www.bp.com/en/global/corporate/energy-economics/energy-outlook-2035.html

Bruns, S., Gross, C., & Stern, D. (2013). *Is there really Granger causality between energy use and output?* https://www.rwth-aachen.de/global/show_document.asp?id=aaaaaaaaaagvwas

CEREN. (2009). In L'Expansion (Nifenecker H.). Le chauffage electrique moins gourmand que le gaz. http://energie.lexpansion.com/habitat/le-chauffage-electrique-moins-gourmand-que-le-gaz_a-39-895.html

DOE. (2017). *Furnaces and boilers*. Department of Energy. https://energy.gov/energysaver/fur naces-and-boilers

DOE/Manufacturing Energy. (2013). *U.S. manufacturing energy use and greenhouse gas emissions analysis*. Department of Energy. https://www.energy.gov/sites/prod/files/2013/11/f4/energy_ use_and_loss_and_emissions_petroleum.pdf

Ecofys. (2011). http://www.ecofys.com/en/info/the-energy-report/

EIA/AEO. (2010). *Annual energy outlook*. http://www.eia.gov/oiaf/aeo/pdf/0383(2010).pdf

EIA/Economy & Electricity. (2013). http://www.eia.gov/todayinenergy/detail.cfm?id=10491

EIA/GDP. (2014). http://www.eia.gov/forecasts/ieo/ieo_tables.cfm

ETC. (2017). *Energy transition commission: Better energy, greater prosperity*. http://www.energy-transitions.org/better-energy-greater-prosperity

Exxon Mobil. (2016). http://corporate.exxonmobil.com/en/energy/energy-outlook

Geohive. (2016). http://www.geohive.com/

Granger, C. W. J. (1969). Investigating causal relation by econometric and cross-sectional method. *Econometrica, 37*, 424–438.

Greenpeace. (2015). http://www.greenpeace.org/international/Global/international/publications/cli mate/2015/Energy-Revolution-2015-Full.pdf

Hou, Q. (2009). *The relationship between energy consumption growths and economic growth in China*. http://citeseerx.ist.psu.edu/viewdoc/download?doi=10.1.1.670.9724&rep=rep1& type=pdf

IIASA. (2012). *Global energy assessment*. International Institute for Applied System Analysis. http://www.globalenergyassessment.org/

JCA. (2018). *Energy consumption for cement production*. Japan Cement Association. http://www. jcassoc.or.jp/cement/2eng/e_01a.html

Klran, B., Guris, B. (2009). *Relationship between electricity consumption and GDP in Turkey*. http://businessperspectives.org/journals_free/ppm/2009/PPM_EN_2009_01c_Kiran.pdf

Mills, M. (2013). *Cloud begins with coal*. https://www.tech-pundit.com/wpcontent/uploads/2013/ 07/Cloud_Begins_With_Coal.pdf

Observatoire de l'energie. (2004). 20 ans de chauffage dans les résidences principales en France de 1982 à 2002. http://www.statistiques.developpement-durable.gouv.fr/fileadmin/documents/ Themes/Energies_et_climat/Consommations_par_secteur/Residentiel-tertiaire/20%20ans%20d e%20chauffage%20res%20principales%20France%201982%20a%202002.pdf

Our World in Data. (2016). http://ourworldindata.org/data/population-growth-vital-statistics/world-population-growth/

Planetoscope. (2016). http://www.planetoscope.com/natalite/5-.html

Schneider Electric. (2014). http://www.schneiderelectric.com/solutions/ww/en/seg/4663977-build ings/4872918-real-estate-office-buildings

Springer. (2012). *Benchmarking energy use in the paper industry: A benchmarking study on process unit level*. https://link.springer.com/article/10.1007/s12053-012-9163-9

TUC. (2012). *Mining resource engineering. Production of aluminum*. Technical University of Crete. https://www.slideshare.net/aparaschos/production-of-aluminum-emphasis-on-energy-and-materials-requirements

UN. (2014). United Nations. http://esa.un.org/unpd/wup/Publications/Files/WUP2014-Report.pdf

Chapter 3
The Transition to a New Power Mix

To sustain this growth in the demand for electricity, new power generation capacities need to be installed and put in operation. Traditionally, fossil fuels have been the main energy source for electricity production. Over the past decade, other sources of energy have emerged, mainly based on renewable sources. They first expanded thanks to government regulations and are now becoming increasingly competitive, leading to a progressive evolution of the electricity mix.

3.1 New Competition Landscape for Electricity Generation

Electric power can be generated using different sources. Traditional generation uses the heat produced by the combustion of a fossil fuel (coal, natural gas, or oil) to pressure up a gas which mechanically rotates a turbine. The rotation of the turbine within an alternator generates electricity. Nuclear power operates under the same principle. Instead of the combustion of a fossil fuel, it is the heat produced by the fission of uranium. Under the same principle, geothermal heat or the heat generated by the sun (in concentrated solar plants) or by the combustion of biomass can also be used to heat up a gas and activate the rotation of the turbines. Hydroelectric plants use the potential energy of the accumulated water which falls in a dam and rotates the turbines. Wind turbines (both onshore and offshore) are rotated by the energy of the wind. Submarine turbines can also use the ocean's current as a source of energy. Finally, photovoltaic solar farms use semiconducting materials which produce electricity when irradiated by the sun.

In 2012, 65% of the generation capacity installed used fossil fuels as a primary source of energy for electricity generation. Nuclear energy represented less than 7% of the total worldwide electricity generation capacity. Outside hydropower, renewable energies accounted for less than 10% of the total capacity installed (Fig. 3.1).

The mix has thus historically favored fossil fuels to the detriment of other solutions, essentially for cost reasons. New variable renewable electricity solutions

© Springer Nature Switzerland AG 2019
V. Petit, *The New World of Utilities*, https://doi.org/10.1007/978-3-030-00187-2_3

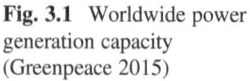

Fig. 3.1 Worldwide power generation capacity (Greenpeace 2015)

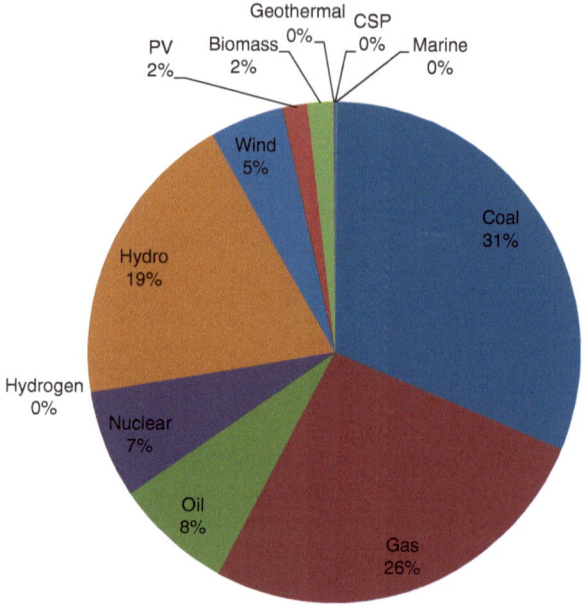

have expanded in the last decade, under the pressure of many governments across the planet, as a response to environmental policies. Today's electricity generation indeed represents around 35% of total carbon emissions. In addition, concerns about the reserves of fossil fuels as well as a too high dependency on fossil fuel exporting countries have led many governments to diversify their sources of energy. Finally, the level of waste in traditional generation averages 60%, since the electricity generating process is extremely inefficient by nature. The combustion of heat, followed by mechanical conversion into electricity through the alternator leads to massive waste of energy in the process. Renewable electricity offers perfect yields as the renewable primary source of energy is free.

These various solutions have different cost profiles, which have evolved dramatically in recent years, leading to a new competition landscape for power generation.

3.1.1   Competitiveness of Power Generation Sources

With the progressive deregulation of many electricity generation markets across the world, a number of instruments and tools have been created to measure the actual competitiveness of the various sources of power. The main indicator of competitiveness of a power generation source used across the globe today is the Levelized Cost of Electricity (LCOE). The LCOE is a complex calculation which basically outlines the actual cost of electricity at which electricity needs to be sold for a power generation project to be profitable. When the LCOE is above the actual wholesale market price (or the retail price in the case of distributed generation), then the project is not profitable. When it is below the actual market price, then the project is profitable. The simplified formula for the Levelized Cost of Electricity is given below (Green Rhino Energy 2016).

$$LCOE\left(\frac{USD}{kWh}\right) = \frac{\sum_{i=0}^{n} \frac{(Ii+Oi+Fi-TCi)}{(1+r)^i}}{\sum_{i=0}^{n} \frac{(Ei)}{(1+r)^i}}$$

In this formula, "I" corresponds to investment costs, "O" to the sum of operation and maintenance costs (fixed and variables), "F" to fuel costs, "TC" to the sum of the possible tax credits applied, "E" to the energy produced (in kWh), "r" to the weighted average cost of capital (WACC), and "n" to the number of years of the project (lifetime of the plant).

The investment cost, also called overnight cost, is generally expressed in USD/kW. The operation and maintenance costs consolidate the variable operation costs related to the power production output, as well as the necessary fixed costs to maintain the plant. The variable part will depend on the production output, and is thus expressed in USD/kWh, while fixed costs are a fixed amount per year. Tax credits are usually a function of the production output and as such expressed in USD/kWh. The energy produced is expressed in kWh.

3.1.1.1   Levelized Cost of Electricity: Overnight Costs

Overnight costs vary significantly across power generation technologies (Fig. 3.2). Traditionally, those for nuclear and renewable energies are among the highest, although the investment cost of renewable has significantly reduced over the past few years. Several studies have looked at the investment costs for different power

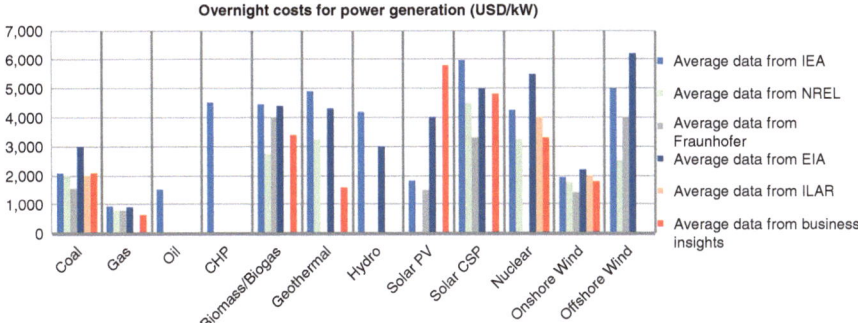

Fig. 3.2 Overnight costs for power generation (Business Insights 2010; EIA/Power Generation 2015; Fraunhofer 2013; © OECD/IEA, Costs 2015; ILAR 2012; NREL 2010)

generation technologies. For this book, data was collected from the Projected Costs of Generating Electricity from the IEA (2015), the National Renewable Energy Laboratory (NREL 2010), the Fraunhofer institute (Fraunhofer 2013), the Energy Information Administration (EIA/Power Generation 2015), the Laboratory on International Law and Regulation (ILAR 2012), and Business Insights (Business Insights 2010).

The data differs from one source to another, although not significantly, with the exception of photovoltaic solar technology, which can be explained by the fact that photovoltaic solar investment costs have considerably decreased over the last few years. As a result, the date of the study is an important input to take into consideration. Older studies (such as Business Insights') present a perspective of photovoltaic solar costs much higher than more recent studies such as that from Fraunhofer. We will see later that solar costs have even further dropped since the date of this study, leading to even lower prices for photovoltaic solar technologies than the lowest ones presented here.

Solar modules represent typically 50–60% of the total costs of the solar farm. The remaining system costs depend on whether the solar farm is a simple rooftop system or a utility-scale system. The costs of modules were divided by five over the 2008–2013 period (© OECD/IEA, Solar 2014; Irena/Solar 2012), mainly thanks to the massive increase in production output from China. The production surge in China drove global production from 5000 units up to 35,000 units per year in 5 years (© OECD/IEA, Solar 2014). System costs have also been divided by three in the 5 years, thanks to productivity improvements, design standardization, and simplification of procedures for grid connection. And these trends have continued, with around 20% decrease year-on-year since (BNEF/Solar 2017) (Fig. 3.3). Consequently, the cost of investment of photovoltaic solar farms has considerably reduced over the past few years, and is projected to continue to do so in the coming 20 years. Other sources of energy have been more stable cost-wise. This has to do with the fact most of them are already mature technologies.

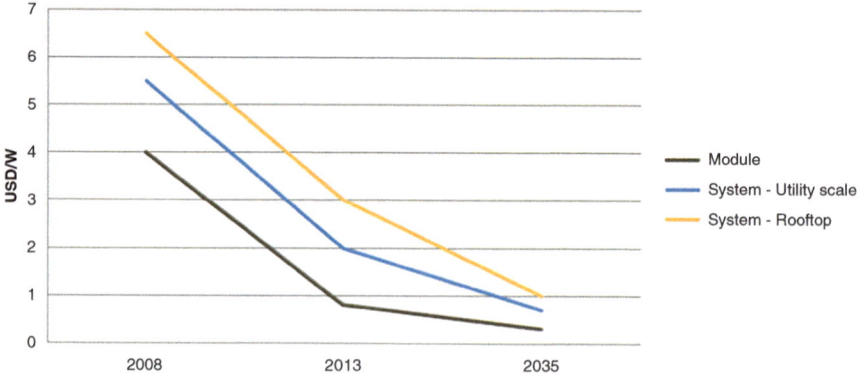

Fig. 3.3 Photovoltaic solar costs evolution (© OECD/IEA, Solar 2014; Irena/Solar 2012)

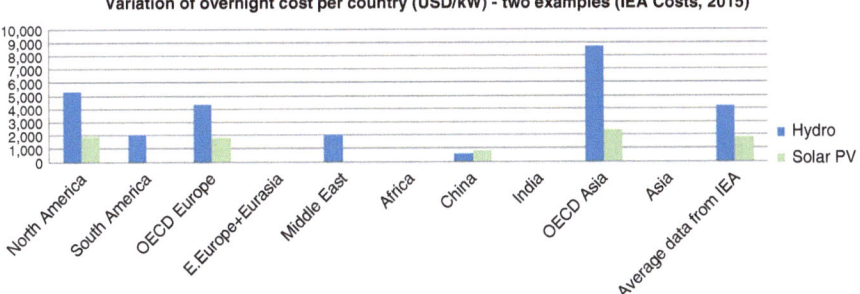

Fig. 3.4 Variation of overnight costs by country (© OECD/IEA, Costs 2015)

The overall cost of construction or overnight costs are typically made of four main components (ILAR 2012). The "bare erection costs" correspond to the cost of construction of building and installation of processes. Then come "engineering, procurement and construction" services costs; they include design, installation, and commissioning services costs. Additionally to these costs, "fees and contingencies" have generally to be borne by the investor. Finally "owners' costs" include pre-production costs, inventory capital, etc. All these costs vary from one country to another; they also depend on the level of experience accumulated (learning curve). Traditionally, "bare erection costs" represent 45–80% of the total overnight costs. They thus strongly influence the final overnight costs. They are also extremely dependent on the country where the power generation units are built.

The perspective of overnight costs is thus highly dependent on the region. Typically, investment costs in OECD countries are much higher than in other regions of the world. Costs of investments in China are among the lowest on the planet. The International Energy Agency (2015) has run an analysis covering different countries. Two examples are provided on Fig. 3.4 for hydroelectric power, with a sample of 28 plants in the Projected Costs of Electricity from the International Energy Agency (2015), and for photovoltaic solar, with a sample of 38 plants. The costs for China are clearly much lower than in other regions of the world. The costs in OECD Asia are among the highest on the planet.

Some sources of power also have different technologies which lead to great variance of overnight costs. For instance, pulverized coal power plants in Korea have overnight costs around 1200USD/W, whereas supercritical coal power plants in Japan or in the Netherlands have overnight costs of above 2500USD/W (© OECD/IEA, Costs 2015). Pulverized coal technology simply uses powdered coal which is fired up to generate thermal energy which can heat up a boiler. Supercritical technologies additionally use very high pressure systems to turn liquid into steam at lower temperatures, leading to lower carbon emissions. The latter technology is more expensive to build. Natural gas power plants can also operate in open cycle (open cycle gas turbines) or in combined cycles. Combined cycles are more expensive to build since they operate with two turbines. Open cycle gas turbines' overnight

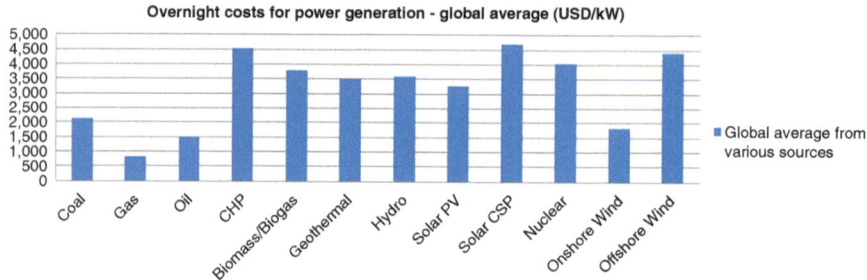

Fig. 3.5 Overnight costs for power generation—global average (© OECD/IEA, Costs 2015)

costs traditionally range around 600USD/W, whereas combined cycle gas turbines' overnight costs range around 1000USD/W.

In summary, all these sources can finally be averaged in order to provide a simplistic perspective of investment costs for power generation (Fig. 3.5). Fossil fuels have lower investment costs than other sources of power, traditionally ranging below 2000USD/kW. Combined Heat and Power (CHP), Concentrated Solar (CSP), Nuclear and Offshore Wind stand out with overnight costs that range above 4000USD/W. Other sources of energy range between 2000 and 4000USD/W. Onshore Wind is the most competitive renewable source of energy, with overnight costs of around 2000USD/W.

3.1.1.2   Levelized Cost of Electricity: Operation, Maintenance and Fuel Costs

Like overnight costs, operation and maintenance costs also vary across various power generation technologies and regions. Operation and maintenance costs are split between fixed and variable costs. Annual fixed costs are set whatever the volume of activity of the power plant. Variable costs will increase as the power output of the plant increases. The Energy Information Administration (2015) has detailed in the United States the split of fixed versus variable operation and main-tenance costs. The ratios are consolidated in the Fig. 3.6. Data for Combined Heat and Power (CHP) technology was retrieved from the University of Chicago (Kalam et al. 2009). Conventional technologies are extremely sensitive to power output. Natural gas power plants have almost no fixed costs, which makes them extremely flexible to operate. Coal power plants are flexible too, although less so than natural gas plants. On the contrary, nuclear power plants have almost half of their operation and maintenance costs as fixed costs, which means they need to be operated a certain amount of time per year as a minimum to amortize those costs. Renewable energies such as wind, solar, or geothermal have no variable costs. Their operation and maintenance costs are entirely fixed. This means that the marginal cost of an additional unit of energy is null.

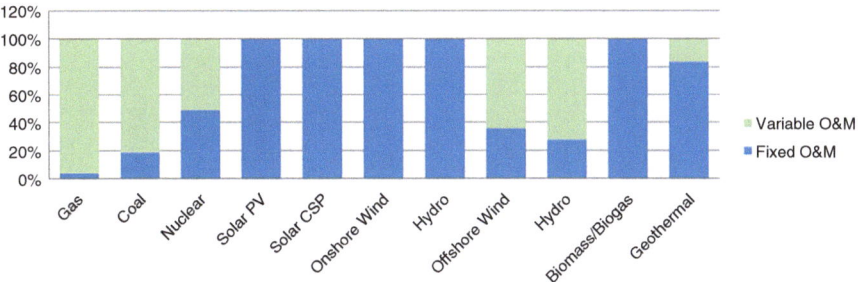

Fig. 3.6 Ratio of fixed vs. variable operation and maintenance costs (EIA/Power Generation 2015; Kalam et al. 2009)

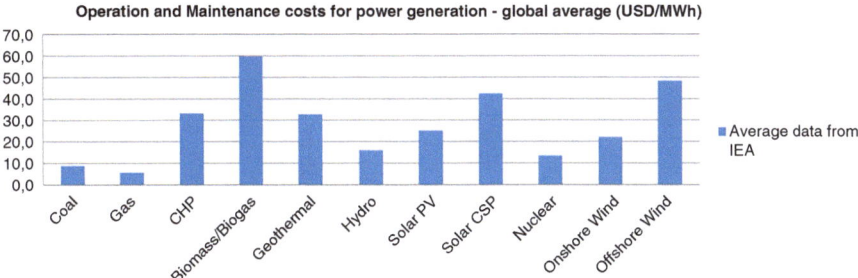

Fig. 3.7 Operation and maintenance costs for power generation—global average (© OECD/IEA, Costs 2015)

Operation and maintenance costs thus vary across technologies in absolute value terms. Conventional technologies have low operation and maintenance costs in general, while renewable energies have higher levels of costs. Biomass in particular has very high operation and maintenance costs, with a global average which ranges around 60USD/MWh, compared to natural gas, which comes in at below 6USD/MWh (© OECD/IEA, Costs 2015) (Fig. 3.7).

Operation and maintenance costs also vary from one region to another, depending on a multiplicity of factors, such as cost of labor, manufactured components, raw materials, logistics, etc. As an example, the operation and maintenance costs for a hydropower plant can range from 5USD/MWh (North America) to 22.6USD/MWh (Japan), giving an average of 16USD/MWh (© OECD/IEA, Costs 2015). Similarly, operation and maintenance costs for photovoltaic solar power range between 8USD/MWh (North America) to 31USD/MWh (Japan and Korea), for an average of 25USD/MWh (© OECD/IEA, Costs 2015). There are therefore strong variations from one region to another (Fig. 3.8).

The final element which determines the operating costs of a power plant technology corresponds to fuel costs, which also vary significantly across different technologies (Fig. 3.9). Fossil fuel based technologies are extremely sensitive to fuel costs. The variability of those on the commodity market is a key element of

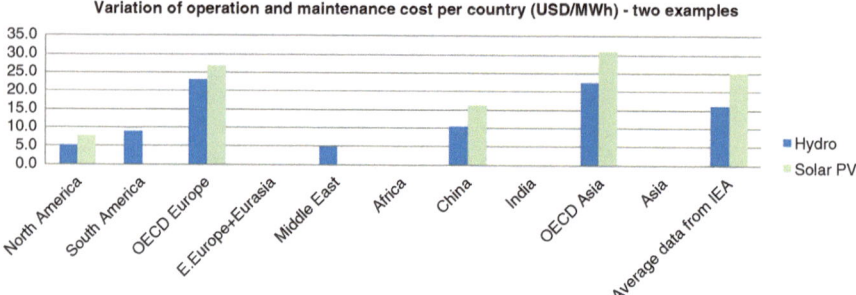

Fig. 3.8 Variation of operation and maintenance costs by country (© OECD/IEA, Costs 2015)

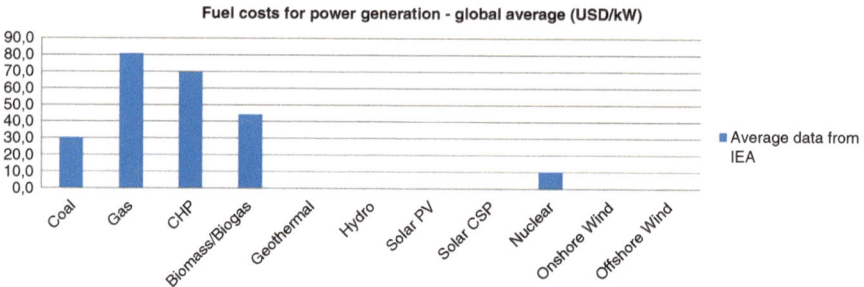

Fig. 3.9 Fuel costs for power generation—global average (© OECD/IEA, Costs 2015)

concern for investors as the amortization period of their investment is usually more than a decade (Vuorinen 2009). Natural gas has the highest level of fuel costs, with around 80USD/MWh as a global average (© OECD/IEA, Costs 2015), followed by Combined Heat and Power (CHP) technology, which ranges around 70USD/MWh, and biomass technology (40USD/MWh). Coal power plants have lower fuel costs (around 30USD/MWh). Nuclear technology has even lower fuel costs at less than 10USD/MWh on average. Finally, renewable energies have no fuel costs, which is a key advantage in terms of ensuring profitability over the project lifetime.

These costs also vary from one region to another, depending on local market conditions. For instance, natural gas in North America costs much less than in other regions of the world, leading to a more competitive situation of natural gas power plant technologies compared to the rest of the world (© OECD/IEA, Costs 2015). According to the same source, the cost of coal is lower in Europe and Africa than in China or in Japan and Korea, leading this technology to be also more competitive in those regions (Fig. 3.10).

When consolidating the figures detailed above, it appears that operation, maintenance and fuel costs vary strongly across the different types of technologies (Fig. 3.11). Generally, renewable technologies have lower costs overall (since they have no fuel costs), and their costs are essentially fixed, a critical factor in the evolution of this new energy landscape. Marginal energy is indeed free.

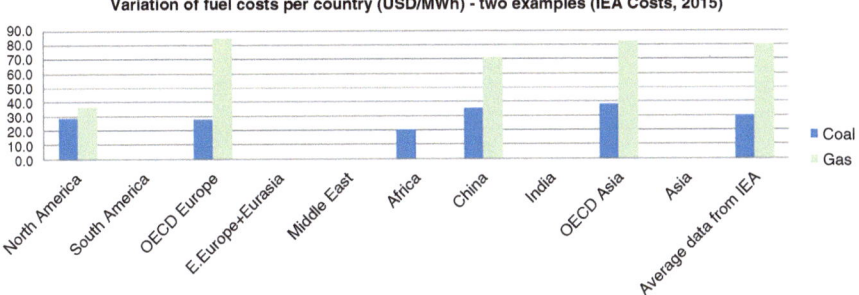

Fig. 3.10 Variation of fuel costs by country (© OECD/IEA, Costs 2015)

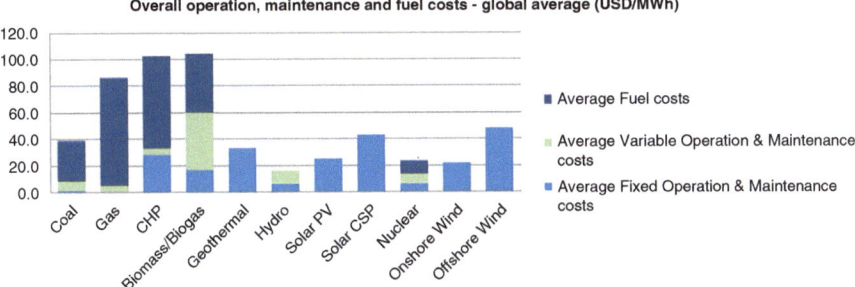

Fig. 3.11 Overall operation, maintenance and fuel costs—global average (© OECD/IEA, Costs 2015)

Conventional technologies such as coal or gas have higher costs overall, but these costs are essentially variable. The profile of their profitability is therefore very much different. On one hand, renewable energies need to amortize their fixed costs (and their investment costs) with significant production output (and as low as possible costs of investments). On the other hand, conventional technologies are much more flexible. The volume of costs they need to bear will vary with their production output. The model can thus accept significant variations of production output.

Finally, as explained earlier, those costs vary significantly from one country to another. The Fig. 3.12 shows the variations from one country to another of coal, natural gas, hydro and photovoltaic solar technologies. Generally, North America benefits from lower operation, maintenance and fuel costs overall. China benefits from lower costs overall for renewable energies, but its costs for coal are actually higher than in Europe.

3.1.1.3   Levelized Cost of Electricity: Capacity Factor

The energy produced by a power generation source depends not only on its nominal power output, but also on its capacity factor, which is the percentage of time it can

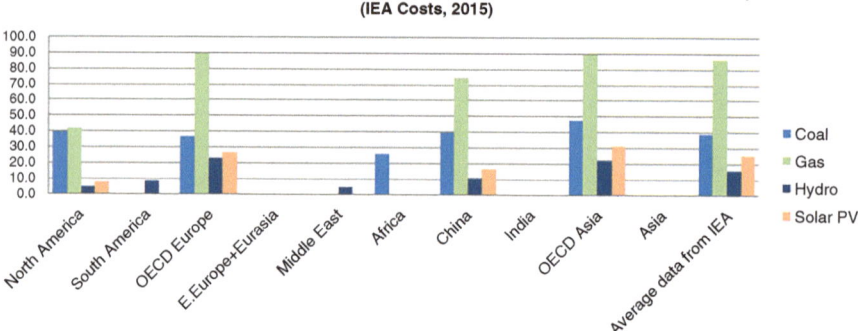

Fig. 3.12 Variation of operation, maintenance and fuel costs by country/region (© OECD/IEA, Costs 2015)

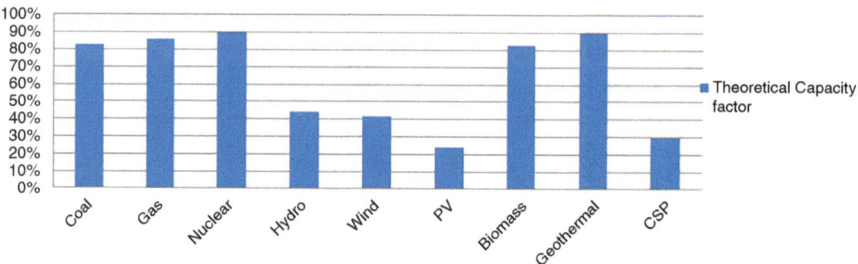

Fig. 3.13 Theoretical capacity factors (IPCC 2011; NREL 2010)

operate. Various technologies have very different capacity factors. Most conventional technologies can technically operate at high capacity factors. Moreover, nuclear technology is obliged to operate almost continuously as the expenses associated with starting up a nuclear plant are extremely high in comparison to operation and maintenance costs. As nuclear technology generates power on the grid practically all the time, it is often referred to as "base load", meaning it supplies the base load that is continuously requested from the grid. Renewable technologies, in particular wind and photovoltaic solar, are variable by nature and thus have a much lower capacity factor.

The capacity factor is first a technical parameter that describes how long a power plant can run over the grid continuously. The National Renewable Energy Laboratory (2010) provides the theoretical capacity factors of each type of power plant technology (Fig. 3.13). The capacity factor for hydropower was retrieved from the Intergovernmental Panel on Climate Change special report on Renewable Energies (2011). Coal and Gas technologies each have a capacity factor of 80% and Nuclear technology even tops 90%. Renewable energies are more variable. Hydropower, Wind, Solar thermal (CSP) and Photovoltaic Solar technologies have capacity factors of well below 40%, in particular Photovoltaic Solar, which barely tops

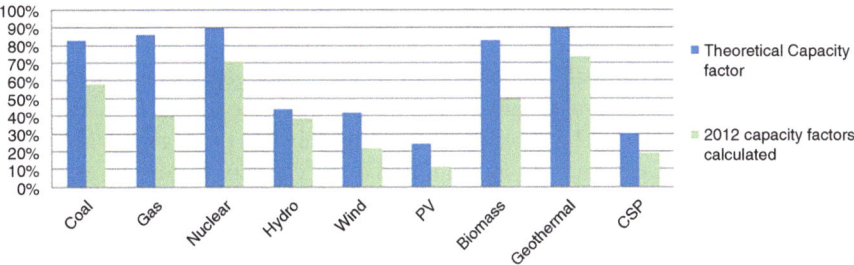

Fig. 3.14   Theoretical vs. measured capacity factors (Greenpeace 2015; IPCC 2011; NREL 2010)

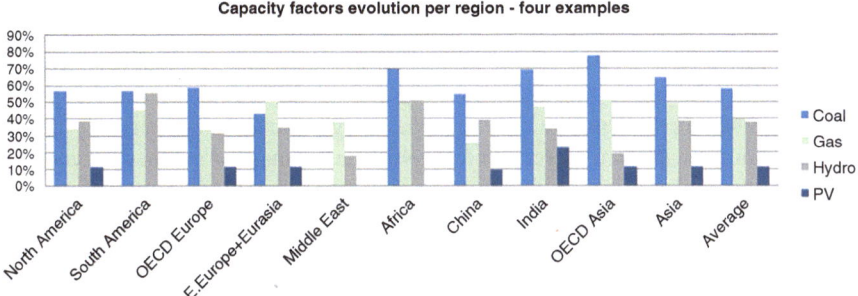

Fig. 3.15   Capacity factors evolution by region (Greenpeace 2015)

20%. Biomass and Geothermal however have much higher capacity factors since they operate under the same principle as conventional power plants. In the case of biomass plants, biomass is substituted for coal or gas, but in essence the process remains the same. In the case of geothermal plants, geothermal energy is continuously available, leading to a high capacity factor.

The actual capacity factors can then be calculated from the actual energy produced divided by the installed production capacity. This enables comparison of the theoretical capacity factor of each technology to the actual performance realized (Fig. 3.14). All technologies obviously perform lower than their potential. Nuclear and geothermal energies reach a capacity factor slightly above 70% on average. Biomass and traditional coal range from 50% to 60% of capacity factor. Natural gas, despite a theoretical capacity factor of 86%, ranges around 40%. This is linked to the fact that these extremely flexible units are often used for peak load management and not as base load. Hydro is close to its theoretical potential, with 39% actual capacity factor worldwide. Wind and solar thermal technologies range around 20% and photovoltaic solar would have a low 10% capacity factor.

Now, these figures are worldwide averages, and this often hides the fact that the performance can be much higher in a number of regions (Fig. 3.15). The performance of coal power plants varies from 43% in Eurasia up to 78% in OECD Asia. The performance of natural gas power plants varies from 26% in China up to 51% in Eurasia, showing as well different types of usages on the grid for similar technologies. The performance of hydroelectric power plants gets as high as 56% (above the global theoretical capacity factor) in South America, a region which heavily depends on hydroelectric power for its electricity generation. Photovoltaic solar is rather balanced across regions, with a capacity factor around 11%, except in India where it tops 23%. The choices of electricity mix in a given region as well as the variation of usages thus strongly influence the capacity factors of electricity markets in a given region. Natural gas is clearly used as "mid peak load" or "peak load" source in China (26% capacity factor), while it is more used as a "base load" or "mid peak load" source in Eurasia (51% capacity factor). Hydropower is used more continuously in South America (56% capacity factor), compared to OECD Asia (19% capacity factor) or in Europe (32% capacity factor), where it is clearly used as a complement to meet extra demand for power.

In summary, variable sources such as wind, solar thermal and photovoltaic solar are generally used as a complement on the network since they are unable to produce continuously. This is one of the critical topics which will be dealt with further (in the Chap. 4) since the cost profile of renewable energies actually fits perfectly that of base load units, with zero marginal costs. Nuclear, biomass or geothermal energies are generally used as a base load source of power. More flexible sources such as coal, natural gas or hydro see their capacity factor adjusted depending on the production plan. This creates an important variety of situations across regions. The economic parameters of the local markets, the history of their power generation footprint, as well as the regulation which applies to those markets have led to various economical and technical situations across regions.

3.1.1.4   Levelized Cost of Electricity: Cost of Capital

The final parameter that influences the LCOE is the weighted average cost of capital (WACC). The WACC depends upon a multitude of factors. Fraunhofer (2013) explains that it depends on the financing structure of the investment. The cost of investment can indeed be funded through equity, which has a specific rate of return (return on equity), and through debt, which has another rate of return (return on debt). The choice of financing has thus a critical impact on the final WACC.

The cost of debt is highly dependent on the country where the debt is taken because of the local government bonds' rating. As an example, the nominal yield of a 10-year bond in Brazil is 12.57% (March 2015), which leads to a real yield of 6.7% once inflation is accounted for, while the nominal yield of a bond in the United States is 2.12% (March 2015), with a real yield around 0.22% (© OECD/IEA, Costs 2015). These rates also evolve depending on the time length of the bond—the shorter the bond, the lower the rate. Additionally, investors taking on a debt will have different ratings depending on their size, their resilience and past performance. A large corporation is likely to have a lower rate than a smaller investor. State-owned electric utilities have typically a high credit rating while shareholder-owned electric utilities traditionally have lower credit ratings.

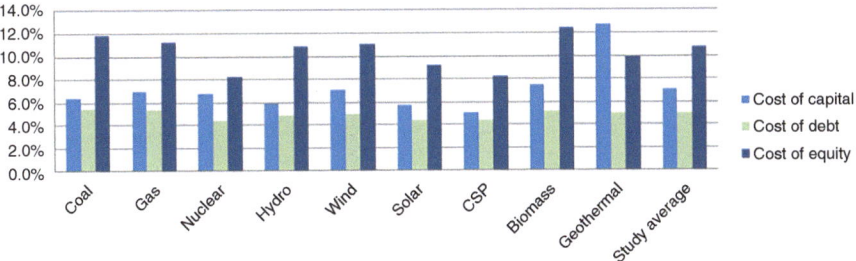

Fig. 3.16 Cost of capital by power technology (© OECD/IEA, Costs 2015)

Corporate equity rates of return also vary from one industry to another and from one company to another. In essence, the market return needs to be higher than a government's bond rate in order to take into account the associated risk. Standard and Poor's (2014) estimated that return has topped 8% in the last 10 years.

Finally, the expected rate of return also depends on the level of risk of the project. Risks can be of different kinds. They can be associated with the regulations within the country, and their possible changes going forward. Environmental regulations represent a risk for traditional fossil fuel based power technologies. Fiscal changes too can affect strongly the profitability of capital-intensive technologies such as nuclear power or renewable energies. Technological risks associated with complex technologies such as nuclear are also taken into account. Finally, external risks such as cost of fuel for fossil fuel based technologies can also impact the expected rate of return.

The International Energy Agency (2015) conducted a survey of different costs of capital for different technologies in different countries (Fig. 3.16). Over 50 power plant investments were surveyed in seven countries. The cost of capital for power plant technologies averaged 7%, with a cost of debt around 5% on average, and a cost of equity close to 11%. The lowest cost of capital applied to solar technologies (both solar thermal or CSP and solar photovoltaic), with a ratio slightly above 5%. The highest cost of capital applied to geothermal energy, topping 12.8%. The cost of debt did not vary significantly across power plant technologies, indicating the relative homogeneity of investors in the power market. The cost of equity was however higher for conventional technologies such as coal and gas, as well as for renewable technologies such as hydropower or biomass. The lowest cost of equity applied to nuclear technology and solar thermal (CSP). Since equity was paid back as dividends, the cost of equity also represented the relative risk of the investment. Nuclear was thus considered a safer bet, compared to coal and gas, which were considered more risky investments in general.

The ratio of debt over the total financing averages 60% (Fig. 3.17). Nuclear is more balanced with 50% of equity financing. This is also due to the large amount of investment required to put a nuclear power plant in operation, as well as an indication of the type of investors in nuclear power, generally public investors. Renewable technologies such as biomass, photovoltaic solar and wind use more

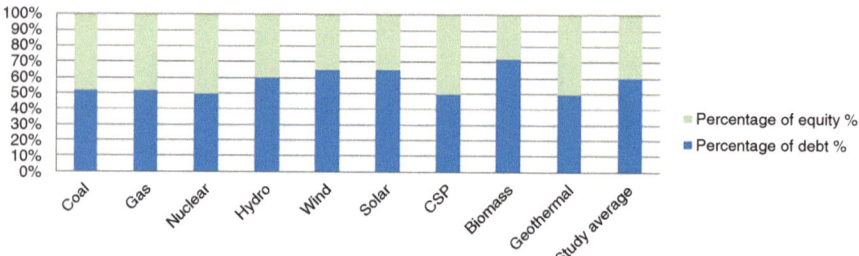

Fig. 3.17 Ratio of financing by power technology (© OECD/IEA, Costs 2015)

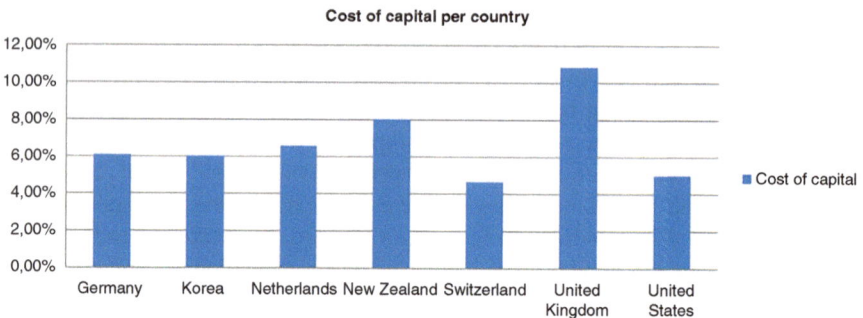

Fig. 3.18 Cost of capital by country (© OECD/IEA, Costs 2015)

debt in the overall financing of their investment, also an indication of the type of investors, with more private investors.

Finally, the WACC will vary in an important manner across different countries (Fig. 3.18). The cost of capital is as low as 4.6% in Switzerland and 5% in the United States. It can however top 10% in the United Kingdom, the average being around 7%.

In the end, the WACC averages 7% for power plant technologies, which corresponds more or less to a normal rate of return in deregulated markets. It varies between the normal rate for government bonds (around 3%) and the typical rate of return for high-risk investments, which generally ranges close to 10%.

All sources will as well have a different sensitivity to the overall cost of capital, depending on the volume of the investment cost in the complete equation. Power generation technologies with a high investment cost with regards to the operation and fuel costs will naturally be more sensitive to it than technologies with a lower investment cost. The Levelized Cost of Capital can be calculated by using the LCOE formula and using only the initial level of investment as an input. This figure gives an idea of how critical the cost of capital is in the overall profitability of a project, as well as its sensitivity, depending on the WACC applied. The Fig. 3.19 shows the estimated Levelized Cost of Capital for each power technology, taking into account a 7% cost of capital. Renewable technologies have traditionally much higher Levelized Cost of Capital than conventional technologies such as coal and

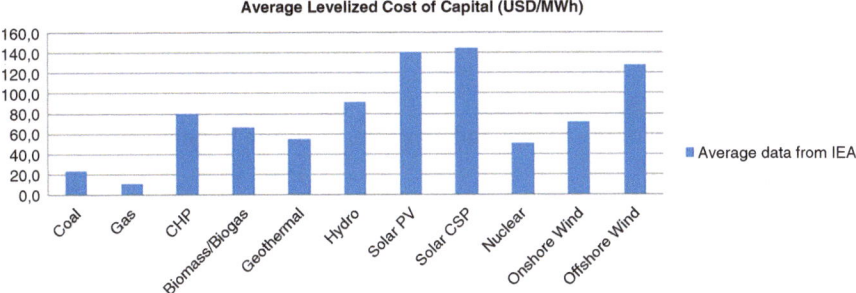

Fig. 3.19 Average levelized cost of capital (© OECD/IEA, Costs 2015)

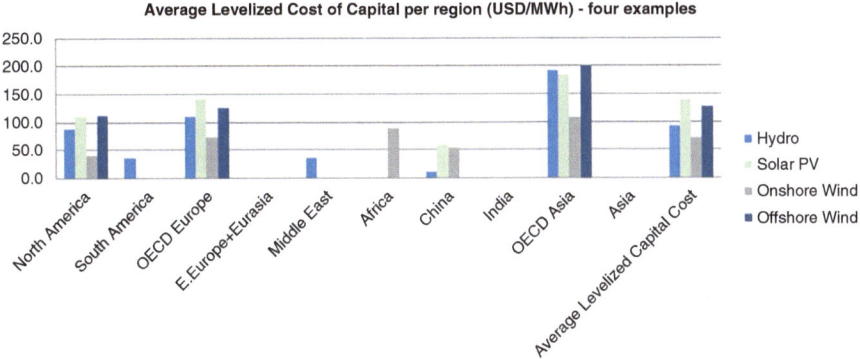

Fig. 3.20 Average levelized cost of capital by region (© OECD/IEA, Costs 2015)

gas, which are not very sensitive to it. Indeed, coal and gas range at or below 20USD/MWh. The necessary price of electricity for solar technologies to be profitable is a minimum of 140USD/MWh, as per the data from the International Energy Agency (2015). Wind is more competitive, with a Levelized Cost of Capital of 70USD/MWh for onshore and 120USD/MWh for offshore. Nuclear, a very capital-intensive technology, reaches 50USD/MWh on average. Capital-intensive technologies are very sensitive to capital costs. A 3% cost of capital indeed yields a Levelized Cost of Capital of 95USD/MWh for photovoltaic solar, 50USD/MWh for wind (onshore), and 24USD/MWh for nuclear, all other things being equal in the 2015 study by the International Energy Agency.

The Levelized Cost of Capital will also strongly vary across regions (Fig. 3.20). Generally, North America and China show much lower Levelized Cost of Capital than OECD Asia or OECD Europe. The cost for hydropower is as low as 36USD/MWh in South America, compared to 191USD/MWh in OECD Asia. The cost for photovoltaic solar is as low as 59USD/MWh in China, compared to 183USD/MWh in OECD Asia, or 141USD/MWh in Europe. The cost of onshore wind is very competitive in North America at a low 40USD/MWh while it tops 108USD/MWh in OECD Asia. The cost of offshore wind is around 110USD/MWh in North America while it tops 200USD/MWh in OECD Asia. Differences are thus extremely important from one region to another.

3.1.1.5   Sensitivity of the Model and Importance of the Regulatory Framework

Many parameters contribute to the determination of the LCOE. Overnight costs have a critical influence on the cost of electricity since they represent the initial investment that an investor has to make to build a power plant. Overnight costs vary across power technologies and countries, as do operation and maintenance costs. Operation and maintenance costs can be of fixed type or variable with the power output. They also have a significant impact on the final cost of electricity. Fuel costs also have a tremendous impact on the cost of electricity and the profitability of an investment, at least for conventional technologies heavily dependent on fossil fuels. The capacity factor, which is the actual number of running hours of a plant, also determine to a large extent the profitability of the investment and the price at which the electricity needs to be sold in order for the investment to break even. Finally, since most of these investments are capital-intensive, the cost of capital (or discount rate) has a critical impact on the profitability of these projects. The investment can rely on debt or on equity. The share of both debt and equity in the overall investment can have a significant impact on the overall profitability of the project, and it obviously varies from one project to another, from one technology to another, as well as from one region to another, depending on the economic situation of the country where the investment is being realized.

Going further, environmental regulations such as setting a price for carbon emissions can have an impact on the profitability of conventional technologies. As well, the lifetime of a plant has an obvious impact on the overall profitability of the investment. The lead time of a project (if late) can also impact the profitability of the investments. When technologies are not mature, or projects are very complex, the lead time can play an important role in the profitability of project.

The International Energy Agency (2015) has studied the sensitivity of the LCOE to these parameters for each power generation technology. Assuming a variation of +/−50% over a given parameter, all other things being equal, the variation of the LCOE was calculated for each technology. The results are presented in the Fig. 3.21.

Most parameters influence the cost of electricity in a symmetrical manner. This means that if the parameter increases by 50%, it has a certain effect on the cost of electricity. If it decreases by 50%, it has a similar effect, but opposite, on the cost of electricity. A few parameters, though, have a non-symmetrical effect. The capacity factor of a power plant technology can have a tremendous negative impact if it falls below a certain level, in particular for renewable technologies, which traditionally operate at low capacity factors. However, above a certain level, the impact becomes minimal. The profitability of a power plant can also be negatively impacted if the lifetime duration of a power plant is reduced. However, the positive impact of a prolongation of the lifetime duration is not extremely important.

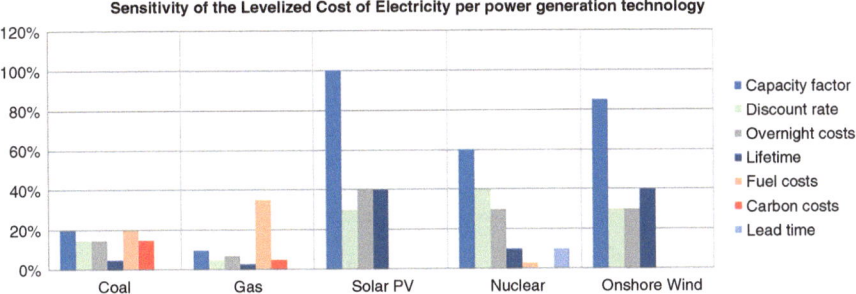

Fig. 3.21   Sensitivity of the LCOE by technology (© OECD/IEA, Costs 2015)

In the end, renewable technologies such as photovoltaic solar and wind, as well as nuclear power, are extremely sensitive to a variation of these parameters. The capacity factor is the most important parameter to take into account. A reduction of 50% of the capacity factor indeed leads costs of electricity to double for photovoltaic solar, to an increase of 85% for wind, and 60% for nuclear. Overnight costs are also extremely important as most of these technologies operate with low operation and maintenance costs and low to nil fuel costs. A variation of 50% of overnight costs leads photovoltaic solar costs to increase by 40%, wind and nuclear to increase by 30%. As a result, the discount rate (or the cost of capital) is key to the overall profitability equation of these investments. They can lead to a cost variation of up to 40% for nuclear power, and 30% for photovoltaic solar and wind. Finally, the lead time can have an impact of around 10% on the LCOE of nuclear power. Fuel costs have no impact on renewable technologies and a very limited impact on nuclear technology (3%).

The situation is very different for conventional technologies such as coal and natural gas. Natural gas power technologies are essentially dependent on the cost of fuel, a key parameter which accounts for almost 80% of the total cost of electricity. A variation of 50% on the cost of fuel can lead to electricity cost variation of around 35%. Coal is more robust a technology and important variations of the different parameters do not lead to variations on the cost of electricity beyond 20%. The capacity factor or fuel costs have a maximum impact of 20% on the cost of electricity in the analysis. Overnight costs and cost of capital do not lead to more than a 15% variation on the cost of electricity. The lifetime of a coal plant has virtually no impact (5%) on the cost of electricity. Now, carbon prices remain an uncertainty which can potentially lift up the cost of electricity of coal technology. They could indeed increase by much more than 50% of their current levels.

According to the World Bank (2015), 40 nations representing 25% of worldwide carbon emissions have put a price on carbon in 2015. The total value of emissions' trading schemes represented 32 billion dollars in 2014. Now, there still exists a vast disparity on how to price carbon. Countries like Mexico, Korea, or Portugal price carbon at a low price below 10USD/ton, while Switzerland and Finland price it at 62USD/ton and Sweden up to 130USD/ton (Fig. 3.22). Beyond governments'

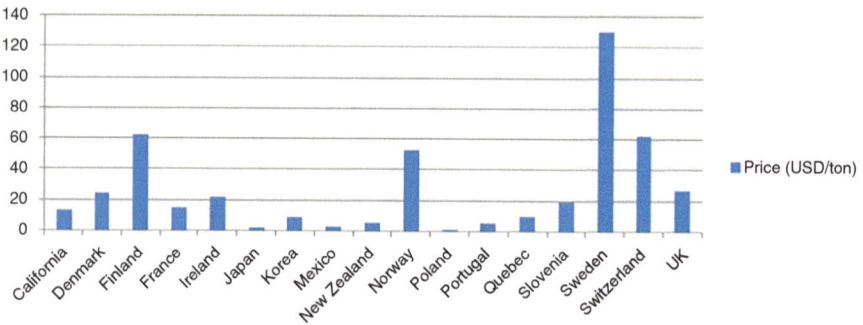

Fig. 3.22 Carbon pricing (World Bank/Carbon pricing 2015)

regulations, the private sector is also moving and implementing internal carbon prices as a tool to direct investments towards low-carbon solutions. Over a thousand private firms have publicly supported such measures in the 2015 New York Climate Summit (World Bank/Carbon pricing 2015). Carbon pricing is thus increasingly becoming a tool for deciding investments, certainly to the detriment of highly emitting technologies, such as conventional power plants, notably coal. The evolution of carbon pricing in the coming years will thus have an important impact on the LCOE of coal and (to a lesser extent) gas power technologies.

Environmental regulations could also push for the adoption of carbon capture storage systems (CCS), which represent an additional investment. Carbon capture storage systems basically separate the CO_2 emitted from the combustion of coal from hydrogen (pre-combustion capture) or steam (post-combustion capture). The separated CO_2 is then compressed and stored, in order to be transportable. The installation of such systems could lead to an increase in the LCOE of up to 90% for coal and 50% for natural gas (Global CCS Institute 2015). The Energy Information Administration (2015) estimates this increase to be more in the range of 30–50%. This increase in the LCOE would mainly come from the increase in the initial capital cost, as well as increased variable maintenance costs.

The LCOE is thus sensitive to a variety of parameters. In this regards, the regulatory framework is critical to ensuring the proper competitiveness of the market and the resulting LCOE.

The LCOE of renewables is also dependent on various regulatory measures which are being or could be put in place to subsidize the installation of new renewable power capacities. This can be done through tax abatements and feed-in tariffs. In over 50 countries, feed-in tariffs have been put in place to encourage the penetration of renewable electricity into the grid (REN21 2013). Considering that the LCOE has been high for renewable energies at the beginning, governments and regulatory authorities have put in place this mechanism to ensure that investors who would invest in renewable power generation would see a reasonable rate of return. They guaranteed, over a certain period of time which could be as long as 15 years, a feed-in tariff, or a price at which electricity was sold, with incentives to maximize the production on the grid. Doing so, they artificially increased the price at which electricity was sold so that these new operators could run profitable investments. As the cost of technology is going down, so are feed-in tariffs, up until renewable technologies reach grid parity, which is the time when renewable power becomes competitive with conventional technologies. The introduction of feed-in tariffs had a double impact on the electricity market. It first tended to increase the overall price at which electricity was purchased, since feed-in tariffs generally correspond to a guaranteed price level. This however did not necessarily reflect on the wholesale market price itself as feed-in tariffs generally take the form of subsidies. Since the production from (and price of) renewable energies was guaranteed, it indeed led to a drop in the electricity price on the remaining units which were competing on "merit order" in the wholesale market, where price is defined by traditional competition between power generation operators and a balance between offer and (remaining) demand. In the end, these subsidies represented in 2013 around 120 billion dollars, which need to be compared to over 500 billion dollars still applying to fossil fuels in many countries (and over 100 billion dollars for the power market alone) (IISD 2016; © OECD/IEA, Subsidies 2016).

Finally, governments can as well influence the LCOE of renewables by working on the actual cost of capital, a key factor of the competitiveness of renewable technologies. First of all, they must strive to provide a transparent regulatory framework which is stable over time, in order to help de-risk investments and ensure investors have access to a lower cost of capital. Second, governments can provide loan guarantees for private investors investing in renewable power generation. This is done at the country level, as in the UK for instance with the Green Investment Bank (UK DECC 2015), or at the international level through the World Bank, the different development banks of Africa or Asia, or the European Bank for Reconstruction and Development (EBRD). These organizations influence greatly the profitability of power generation investments by reducing the cost of capital (essentially the return rate on debt).

In conclusion, power generation technologies are not sensitive in a similar manner to the various parameters which contribute to setting up a LCOE. While conventional technologies are more sensitive to fuel costs, renewable technologies (and nuclear) are more sensitive to the cost of capital as well as the capital itself, and the overall capacity factor. These different profiles of profitability are key to understanding the model under which these generation technologies need to be used on the grid. While renewable and nuclear technologies need to operate at their maximum possible output, with an initial investment as low as possible, coal and gas technologies are more flexible, although very sensitive to the cost of fuel and the various environmental regulations which can modify their profitability profile. In the end, the regulatory framework is key to defining the proper balance of these various technologies. Governments and international institutions play important roles in putting in place the right set of incentives. Regulations around carbon pricing, feed-in tariffs and various abatements, as well as guaranteed loans and investment banks can help set up a certain market balance where all technologies can contribute to what they are eventually purposed to do.

3.1.2   The Upcoming Evolution of the New Power Generation Competitive Landscape

The LCOE outlines the price at which electricity needs to be sold in order for an investment to be profitable. It consolidates various factors such as initial capital and cost of capital, operation and maintenance costs, fuel costs, taxes such as carbon pricing, as well as the energy production forecast as a function of the capacity factor. Most of these factors vary from one project to another, because of the conditions of the project, of the country of destination, of the economic situation of the investor, etc. A global average can however be provided based on the analysis from the International Energy Agency over 180 power plants in the world (© OECD/IEA, Costs 2015). The Fig. 3.23 shows the various calculated LCOE for each type of power plant technology, averaged globally. Geothermal and solar thermal (CSP) are not very well represented with only six and four samples, respectively. The other technologies have in the range of 15 samples throughout the world, with the exception of hydro, which has 28 samples, and photovoltaic solar, which has up to 38 samples. This sampling is thus generally fairly representative of the reality.

The analysis from 2015 shows that nuclear power is the cheapest source of energy, with a LCOE slightly above 70USD/MWh (decommissioning costs are not included) (Fig. 3.23). Other conventional technologies such as coal and gas are also cheaper than any other source of power, even with carbon pricing at its current level. Coal ranges around 80USD/MWh, while gas reaches 100USD/MWh. Combined heat and power technologies (CHP), mostly based on gas and biomass, are the most expensive technologies, around 190USD/MWh, but this figure does not account for heat credits, which range around 50USD/MWh. If they are taken into account, CHP technologies become more affordable (© OECD/IEA, Costs 2015). Some renewable technologies are already extremely affordable, such as geothermal energy, hydropower and onshore wind farms. Offshore wind is more expensive, essentially because of the capital intensity required to put in operation a wind farm offshore, at around 180USD/MWh. Solar remains an expensive solution, with a LCOE around 160USD/MWh for photovoltaic solar, and slightly above

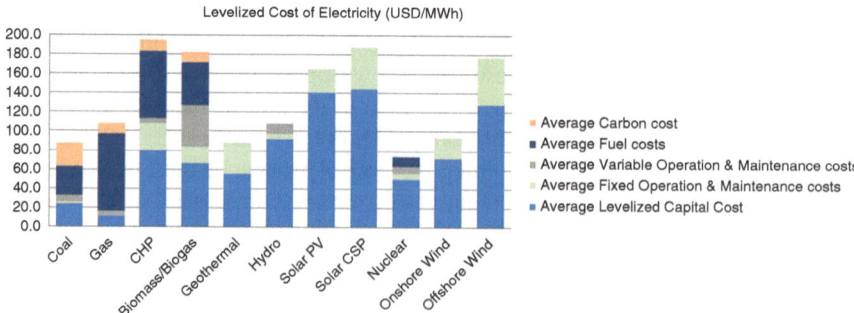

Fig. 3.23   LCOE by technology (Source: author's own calculation based on data from IEA)

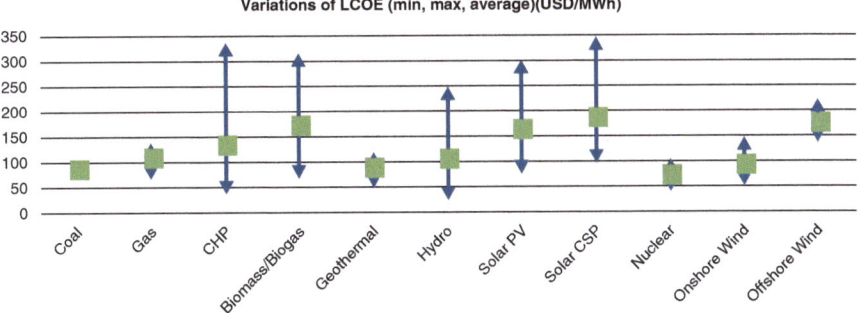

Fig. 3.24 Variations of LCOE by technology (Source: author's own calculation based on data from IEA)

180USD/MWh for solar thermal technologies (CSP). As discussed already, the profiles of these costs vary in essence. While technologies based on coal, gas, and biomass are strongly dependent on the cost of fuel and the uncertainties around carbon pricing, other technologies essentially rely on the initial capital invested and the associated cost of capital, which to a great extent define the final cost of electricity.

These figures are global averages and actually hide the variety of situations which apply across the different regions. Some power technologies have a high variety of LCOE (Fig. 3.24). If the samples for coal, gas, nuclear, geothermal or wind do not vary widely, the costs of electricity for solar technologies, combined heat and power (CHP) and biomass show large differences across countries and technologies. Photovoltaic solar present for instance a large array of LCOE. While costs are around 75USD/MWh in China, they go up to 300USD/MWh in Japan. In Germany, within the same country, costs vary between 127USD/MWh and 223USD/MWh. Hydropower also varies significantly across and within countries. Costs in Brazil vary from 30USD/MWh to 78USD/MWh, and those in Switzerland from 72USD/MWh to 250USD/MWh. Globalized costs must thus be taken with precaution as they hide a strong disparity of local situations which are impossible to depict precisely.

The cost of capital assumed above has been considered fixed at 7% for all sources of power. Now, different technologies offer different risk levels. Conventional technologies such as coal, gas and CHP are mature, but the uncertainties with regard to carbon pricing and other environmental regulations make those investments more risky by nature. Nuclear power is not concerned by carbon pricing, but the risk level associated with these large construction projects must also be taken into consideration. Renewable energies are much less risky, especially in the current framework of feed-in tariffs and sometimes guaranteed loans from public investment banks. Therefore, a number of parameters could be changed to fine tune the comparison above. First, the cost of capital could be considered different across the various technologies. To a certain extent, conventional technologies such as coal, gas and combined heat and power (CHP), as well as nuclear power, could be considered risky investments in the current regulatory context. A premium could be applied to their cost of capital.

Carbon pricing is today an emerging concept. It generated in 2014 around 32 billion dollars of revenue, for around 7 gigatons of CO_2. This represents a worldwide average carbon price of 4.57USD/ton. This price is expected to increase in the coming years as many economies as well as the private sector progressively buy into the concept of setting a price on carbon. The European Union has published a study from Thomson Reuters (2014) in which carbon price is expected to be lifted up from 10USD/ton on average today in Europe to up to 50USD/ton by 2030. Synapse Energy (2016) published an outlook on carbon pricing forecast for the United States, where carbon pricing is not applied in all states, and estimated that the price could go up to 40USD/ton by 2035 (midrange). Carbon pricing is thus projected to expand across regions and to increase in value.

Finally, LCOEs are expected to evolve over time. As explained above, the costs of photovoltaic solar, following the traditional learning curve, have dropped dramatically in recent years. Forecasts show that this trend should continue (© OECD/ IEA, Solar 2014). Actually, most recent studies (less exhaustive as the International Energy Agency one used here) have shown that this drop could come at a faster pace than originally anticipated. In 2017 for instance, BNEF estimated that the average cost of utility-scale solar had already reached 80USD/MWh in China and Germany, and 60USD/MWh in the US. Additionally, some projects were identified with LCOEs close to 30USD/MWh under specific conditions (BNEF/Solar, 2017). Those LCOE are already significantly lower than the costs projected in the 2015 study above. The cost for wind technologies have also moved downwards over the last decade, and should continue to do so, albeit at a slower pace than solar technology. Other technologies, being more mature, should not see such improvements. On the contrary, heavier regulation of nuclear or conventional technologies should lead to increased capital costs as well as increased operation and maintenance costs.

The Energy Information Administration (EIA/Power Generation 2015) has forecasted the evolution of LCOE between 2020 and 2040 for various power technologies in the United States (Fig. 3.25). The analysis shows that the cost of natural gas,

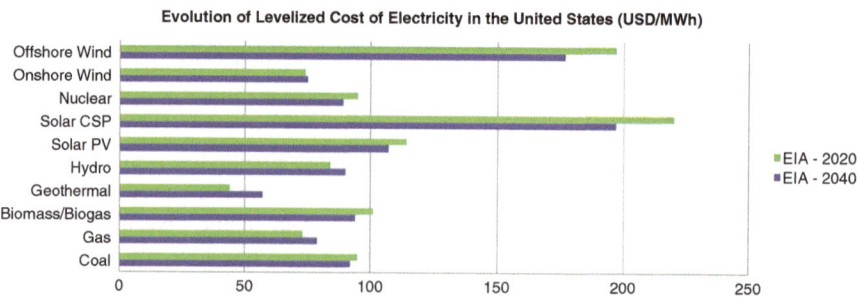

Fig. 3.25 Evolution of LCOE in the United States (EIA/Power Generation 2015)

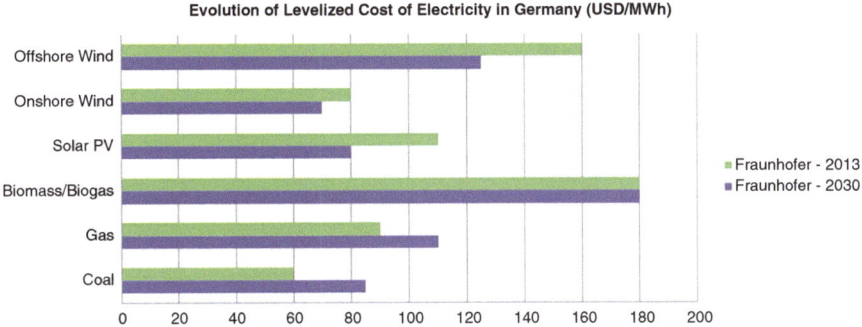

Fig. 3.26 Evolution of LCOE in Germany (Fraunhofer 2013)

geothermal and hydro technologies should slightly evolve upwards, while coal and onshore wind should remain stable. The cost of solar technologies, offshore wind, biomass and nuclear should evolve downwards. According to the projections from the agency, there should however not be any change of paradigm in LCOE in the United States in the coming years.

Fraunhofer (2013) is more radical for Germany. It expects a strong increase in LCOE for coal and gas, with a strong reduction of the costs for photovoltaic solar and wind technologies, leading to a change of paradigm where both solar and wind technologies would break even with conventional technologies by 2030 (Fig. 3.26).

Other sources present different pictures (NREL/Projections 2015; Greenpeace 2015; PSI 2014). All sources acknowledge a drop in costs of wind by an average of 15% for onshore and 24% for offshore. Photovoltaic solar is expected to drop the most, by an average of 40–50%. Recent data points (BNEF/Solar, 2017) suggest that the drop could be even larger and research could take it even further, notably with the emergence of new materials (nanotechnologies, perovskite) or new technologies (leveraging infrared, etc.). Solar thermal (not represented in the graph) could drop by an average of 37%. Other renewable technologies (not represented either) should remain stable as they are already mature. Conventional technologies such as coal and gas should see their costs increase by an average of 14% for coal and 21% for gas, while the cost of nuclear should slightly increase by 10% on average. We see from Fig. 3.27 that the projections from the Energy Information Administration remain very conservative compared to other sources, such as Greenpeace or Switzerland's Paul Sherrer Institut (PSI) (Fig. 3.27).

The actual LCOE must thus be taken carefully as many parameters are expected to evolve. Figure 3.28 computes the actual LCOE for different power generation technologies with a number of assumptions made on the actual evolution of some key parameters. First, the cost of capital is lifted up to 10% for coal, natural gas, combined heat and power and nuclear, in order to take into account the inherent risk factor associated with these technologies in the given context of environment protection. The cost of capital for photovoltaic solar and wind technologies is lowered to 3%, taking into account the highly favorable system in place today to

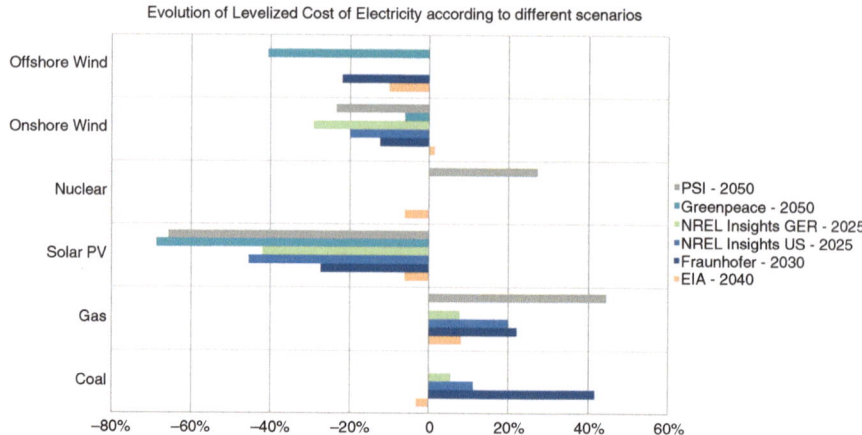

Fig. 3.27   Evolution of LCOE (EIA/Power Generation 2015; Fraunhofer 2013; Greenpeace 2015; NREL/Projections 2015; PSI 2014)

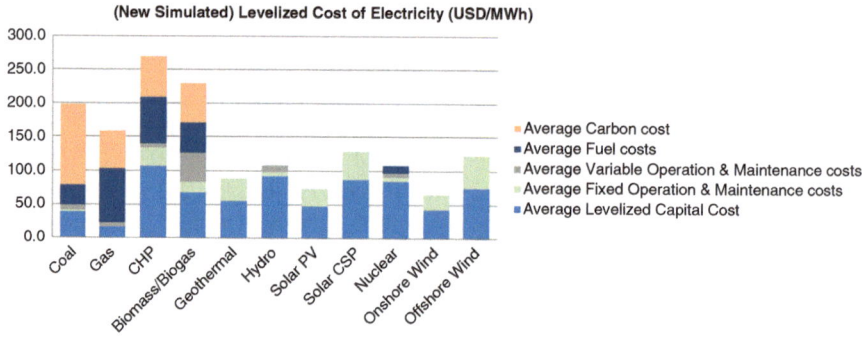

Fig. 3.28   New simulated LCOE (Source: author's own calculation)

promote their development. As well carbon price is multiplied by five from the original analysis from the International Energy Agency (2015), to around 25USD/ton. Finally, the cost of investment is revisited taking into account a drop for wind technologies of 15%, a drop for solar technologies of 50%, and an increase for coal and gas of 15% and for nuclear of 10%.

The results are striking as they show that renewable energies become far more competitive than conventional technologies. Nuclear power becomes more expensive, yet remains profitable. Cost-wise, conventional technologies are increasing dramatically, in particular coal and gas. Coal is becoming one of the most expensive sources of power, with a LCOE of 200USD/MWh, and gas is reaching around 150USD/MWh. In the frame of the assumptions above, the power market experiences a massive change of paradigm with renewable energies substituting all conventional technologies from a competitiveness standpoint. As already explained, most recent studies (BNEF 2017) show that these photovoltaic solar price points

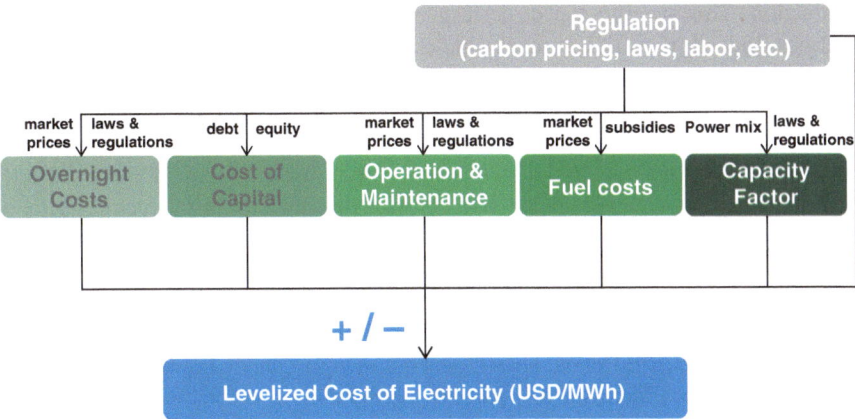

Fig. 3.29 A model for the local power market competitiveness

could have already been reached in a number of regions, and sometimes exceeded in a few projects, leading to the belief that the trend described above could happen at a much faster pace, solar being a true cost disruptor. The above results must thus be taken with caution, as most of the data are from an exhaustive analysis from 2015. Since then, true price shifts have continued to occur. The interest of this analysis is essentially to detail the factors that impact LCOE levels for different technologies and demonstrate the key trend towards mass adoption of competitive renewable energies.

To summarize, the LCOE is a complex concept to comprehend, which may lead to wrong interpretations. Designed to be an indicator of the economic competitiveness of different power technologies, its calculation is based on many assumptions which, provided they are modified by the influence of policies and regulation, can drastically modify the balance presented. A global model can however be established to understand the global and local dynamics of LCOE calculation and therefore the evolution of the competitive landscape of power generation in a particular region (Fig. 3.29).

If conventional technologies are today in most cases more economical than renewable electricity, it is clear that this situation is not meant to last. The evolution of overnight costs for renewable energies, the various environmental regulations as well as their consequences on the cost of capital will deeply modify the competitive balance between traditional technologies and renewable energies. Renewable energies are set to become the competitive choice for power generation in the coming years. A massive transition is thus expected to play out over the coming decades. Renewable energies' cost structure corresponds to that of base load units. Eventually, renewable energies will replace to a large extent conventional base load units, in the sense that they must be privileged to operate. One of the key issues that will arise is the inherent intermittency of their operation, which leads to a number of grid issues. These issues will be reviewed in the next chapter. In the long term, the specificity of renewable energies also lies in the fact that it has zero marginal cost of production. This specific aspect corresponds to the unique promise of renewable energies. Indeed, if sufficient renewable energy capacities are installed, the cost of electrical energy virtually becomes free. Spot markets in Germany have already demonstrated this several times, with prices even going negative at certain times of the day. One of the main promises of renewable energies is thus free electrical energy. In an all-renewable world of energy, the cost of electrical energy would become nil and consumers would pay only for the capacity infrastructure. This is quite similar to what the telecom industry experienced at the beginning of the twenty-first century. When one thinks of the benefits of "free" Internet access (think the multitude of applications which have changed our lives), the development of a "free" energy system would potentially bear spectacular consequences. The transition towards a different power generation mix cannot however be realized without financing. Financing will partially be done by investors but the cost of the transition will eventually be borne by consumers. This can be done both through a well-designed increase in retail price or through various subsidies financed by taxes. In mature economies, this would help speed up the substitution of traditional technologies reaching end of life with renewable technologies. In new economies, the right setup of incentives would help new capacities to be built to be renewable in order to accelerate the change of mix.

3.1.3   The Transition Leads to a Pressure on Prices

3.1.3.1   A Regional Perspective

Different regions experience a variety of local situations (Figs. 3.30, 3.31, 3.32 and 3.33). The current electricity generation mix is more or less diversified, and the

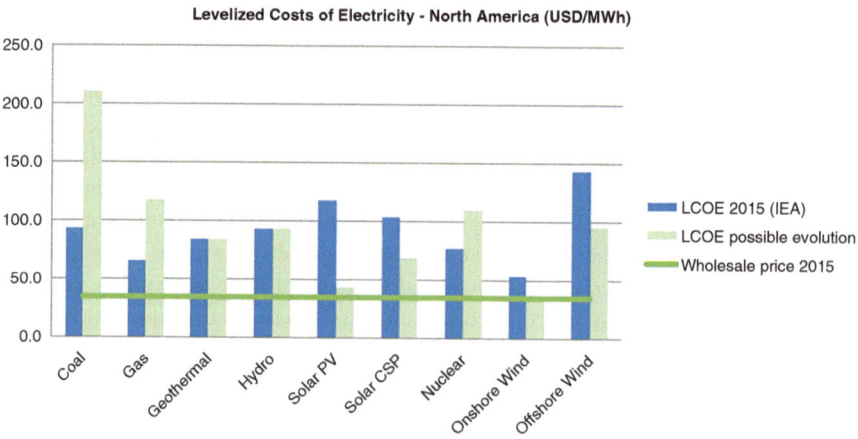

Fig. 3.30 North America LCOE (EIA/Wholesale 2016; © OECD/IEA, Costs 2015)

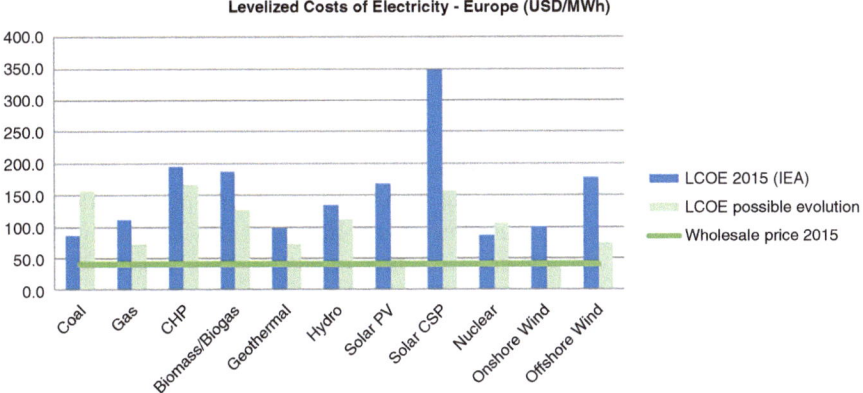

Fig. 3.31 Europe LCOE (EIA/Wholesale 2016; © OECD/IEA, Costs 2015)

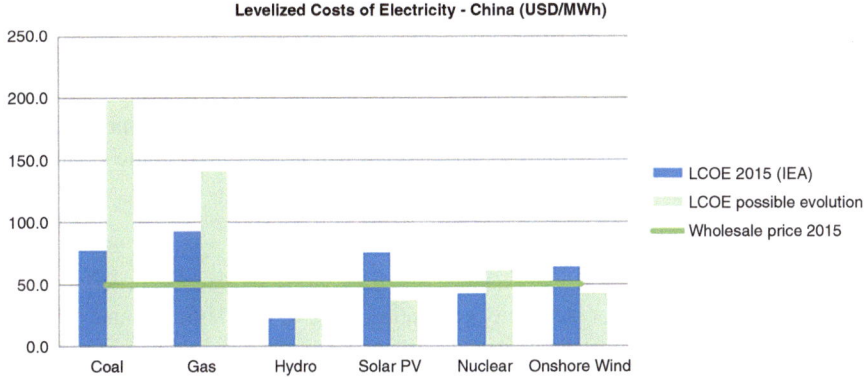

Fig. 3.32 China LCOE (EIA/Wholesale 2016; © OECD/IEA, Costs 2015)

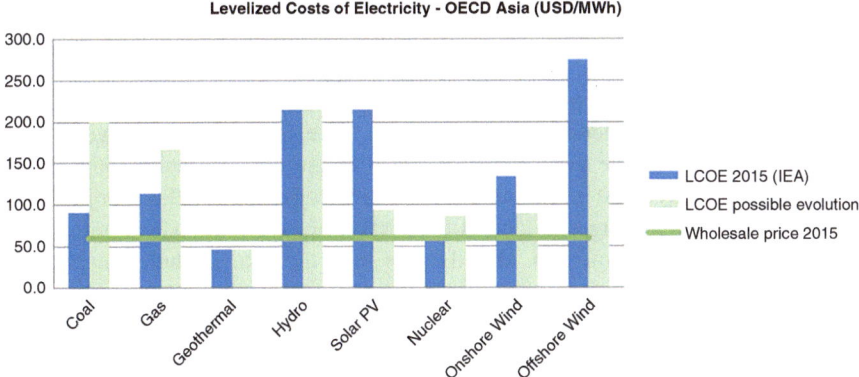

Fig. 3.33 OECD Asia LCOE (EIA/Wholesale 2016; © OECD/IEA, Costs 2015)

competitive landscape more or less reflects this situation. Ahead of the challenges or opportunities that apply to the different power generation technologies, each region will have to adapt accordingly and follow its own different path.

Typically, most regions (except South America and, to a lesser extent, Europe) are extremely dependent on fossil fuel technologies. Evolutions of technology costs should impact all these regions heavily. Renewable energies represent a valid pathway to all regions of the world, but with different starting points. Grid parity is closer in North America and China than it is in Europe or OECD Asia. In OECD Asia, nuclear remains a very competitive source of energy. This is also the case in China, and remains an open question in Europe, in particular France.

Finally, the level of the wholesale price is very low in North America and Europe compared to new units' LCOEs because it is based on already amortized generation units, leading to little incentive to invest further in substitutions. Subsidies are thus playing a critical role in forcing the substitution faster than the normal rate of existing capacity retirements. In other regions, the growth of demand and the relative wholesale price level leads to lower issues on that front.

In conclusion, renewable energies are expected to take a significant share of the world electricity generation mix and, to a lesser extent, nuclear power (in certain regions). Renewable energies will primarily complement and substitute traditional base load generation (because of their cost structure), leading to a number of grid management and market operation issues, which will be reviewed in the Chap. 4.

3.1.3.2   Overall Pressure on Prices

The various projections for electricity generation mix can help simulate an evolution of the average cost of electricity in the coming years. This is obviously a highly theoretical exercise, considering the relative lack of data, the inaccuracy of the projections, and the fact that we are using global averages. Nevertheless, it shows undoubtedly that an adjustment to prices is needed, in particular in mature economies. The LCOE of new generation units is higher in most cases and in most regions than the actual wholesale price, an indication that existing generation units are more competitive than new units. The Institute for Energy Research (2015) has analyzed the evolution of the LCOE over time for existing generation resources. It has clearly shown that the LCOE tends to decrease over time for conventional generation resources because of the progressive decrease of the cost of capital (Fig. 3.34). Basically, once the capital is reimbursed, the LCOE corresponds to the marginal cost of production, which ranges around 30USD/MWh for conventional resources (the example is in North America), slightly less than the actual wholesale price of electricity.

As a result, the revamping of existing generation units reaching end of life in mature countries, or the support of investments in new power generation units to meet the increase in demand in new economies, will lead to a necessary increase in the wholesale price (without subsidies). This is particularly true in North America and Europe markets. In other regions, wholesale prices are closer to the LCOE of

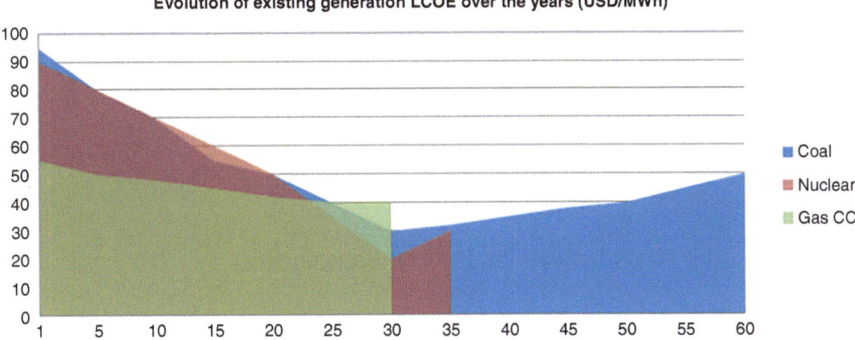

Fig. 3.34 Evolution of existing generation LCOE over the years (Institute for Energy Research 2015)

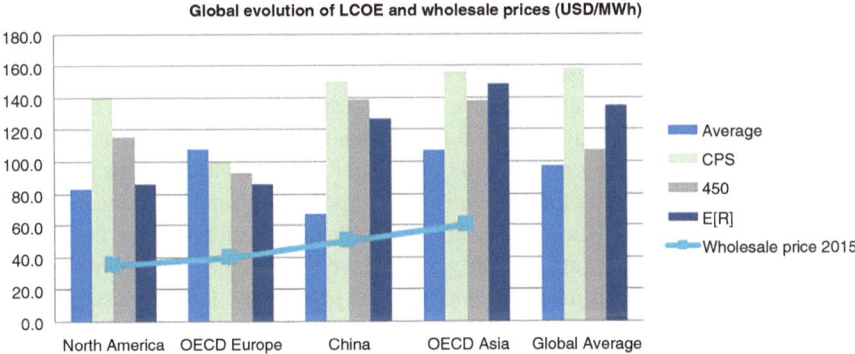

Fig. 3.35 Global evolution of LCOE and wholesale prices (Source: author's own calculation based on data from IEA and Greenpeace)

new generation units. On top of this, the choice of energy mix (move away from fossil, decision on nuclear, increase renewable, etc.) will influence the average electricity price which will require to be reached for this transition to be made possible. The choice to move to renewable makes perfect sense in North America and China. In Europe and OECD Asia, nuclear remains an important and competitive source of energy as well to keep prices low. The evolution of the LCOE itself over time, because of the changes in the external environment of constraints applying to such projects, will also greatly influence the evolution of the cost of electricity, with a definite trend upwards. Now, the evolution of LCOE of photovoltaic solar could to a large extent mitigate this as it reaches new low levels in the coming decades.

In the end, the average wholesale price of electricity (without subsidies) is likely to increase in the decade to come to meet the actual LCOE of new generation units and their actual evolution (Fig. 3.35). Quite paradoxically though, various subsidies and the general regulatory framework could prevent this from happening. They could even lead to the reverse situation. Indeed, if they help offset (at least partially) the effect of the cost of capital, then the marginal cost of variable renewable solutions could even lead to a drop in electricity prices. This is in essence what is being witnessed today in some European countries. Market regulation will play a key role to properly tune the market to help achieve the transition progressively.

3.2 Massive Investments in Power Generation

3.2.1 Various Scenarios of Power Generation Investments by Source

Two main trends are reshaping the market.

First, electricity consumption is expected to increase significantly (between 53% and 85%, depending on the projections) in the coming 20 years. A large share of the already "locked-in" increase is in new economies (around 80%) as economic development of once isolated regions and rural areas turn them into cities and economic centers. A fantastic potential for growth also exists in mature economies because of the progressive switching of traditional fuels to electricity. This transition has already been happening for 20 years and is expected to continue in the next two decades. It could as well accelerate, notably with the development of the digital economy, proving the current forecasts wrong. Nevertheless, the current evolution of electricity consumption in mature economies has been in the last years stagnant at best, if not negative, due to the combined effects of the 2008 economic crisis and the impact of energy efficiency measures.

Second, renewable energies are progressively emerging as a valid option for power generation. These technologies are getting more economically competitive as the scale of production increases and as they progress along the learning curve. As well, the various regulatory measures taken across countries favor the emergence of renewable power as governments attempt to drastically cut carbon emissions of the power sector, which account for 35% of worldwide emissions.

These various changes lead to a different set of scenarios regarding power generation, depending on how fast these elements combine together.

The worldwide power generation capacity in 2012 amounted to 5600GW for all sources of power. Various scenarios point to a significant increase in power generation within the coming 20 years. The scenarios from the International Energy Agency (2014) range between 8500GW and 9100GW for 2030. The high scenario corresponds to the CPS scenario, which assumes no changes happen in the current regulatory framework. The NPS scenario, which assumes the current changes discussed will be implemented, and the 450 Scenario, which assumes additional regulatory measures will be implemented to limit the greenhouse gas concentration to 450 ppm, both lead to a projection of around 8500GW. Finally, the Energy [R] evolution scenario (Greenpeace 2015) forecasts an increase in the power generation capacity up to 11,500GW. Other scenarios not presented here back up these ones. Statista (2016) projects a growth of capacity up to 10,800GW; Deloitte (Deloitte/ Capacity, 2015) up to 11,000GW; the World Energy Council (2013) indicates a growth up to 9800GW. Only the Energy Information Administration (EIA/Capacity, 2016) has a slightly lower forecast of 7500GW.

These scenarios present a great variety of power generation mix (Fig. 3.34). The CPS scenario assumes a continuous increase in conventional fossil fuels to meet the increased demand. Coal based power capacity is expected to grow by 57%, with an increase of over 1000GW over the period. Natural gas is also expected to grow by over 60%, or close to 1000GW. Oil would however decrease by around 135GW, which corresponds to 31% of the installed capacity. Nuclear energy is expected in the scenario to grow by 26% with an increased 100GW of capacity. Renewable energies would grow significantly over the period, with 41% for hydropower (446GW), 200% for wind (500GW), and 400% for solar (400GW). Biomass, geothermal energy, solar thermal (CSP) and marine would also more than double, and grow by a cumulated 155GW. In this scenario, fossil fuels would then represent half of the increased power generation capacity.

The NPS scenario follows a similar trend although the share of fossil fuels is lower than in the previous scenario. Fossil fuels would indeed represent only a third of the total power generation capacity increase. Coal growth would be halved, around 500GW. Natural gas would grow by only 50% around 700GW. Oil would decrease more significantly. In this scenario, nuclear energy would grow in a more significant manner, around 150GW or 40%. Hydropower would grow by a similar amount as that of the previous scenario, around 480GW. Wind would grow by around 640GW (against 500GW in the previous scenario). Solar would grow by a similar factor as in the CPS scenario. The remaining power would be supplied by other renewable sources, such as geothermal, biomass, solar thermal and marine energies, which would altogether grow by almost 200GW.

In the 450 scenario, fossil fuels would remain stable, with a low decrease of 50GW of capacity. While natural gas would keep growing by around 500GW (or 36%), coal would drop by almost 400GW (−22%) and oil by more than 200GW (−50%). Nuclear energy would grow by more than 80%, contributing more than 300GW of additional capacity. All renewable energies would grow at a faster pace as well. Hydropower would be lifted up to 640GW (against 480GW in previous scenarios). Wind power would increase significantly by 1,000GW (or almost four times the current capacity). Photovoltaic solar would grow by close to seven times, with a cumulated contribution of around 620GW of additional capacity. The contribution of other sources would also double, around 330GW, notably thanks to biomass (177GW of additional capacity) and solar thermal, which additional capacities would reach 100GW, compared to 38GW in the NPS scenario.

Finally, the Energy [R]evolution scenario shows a much sharper decline of coal and oil, one partially compensated by natural gas. Oil would be almost phased out, with an over 70% decline (around 300GW). In this scenario, there would not be any increase in the power generation capacity of fossil fuel based technologies. Nuclear energy would almost be phased out with over 300GW of decline (−80% of current capacity). These evolutions on conventional technologies would be compensated by a sharp increase in renewable energies. Due to their variable nature and therefore low capacity factor, the corresponding increase in renewable capacities would thus end up being much larger than in previous scenarios, leading to a total increase of 5800GW of power capacity, compared to 3000–3400GW in the other scenarios. In this scenario, hydropower would not increase by more than 30% to around 300GW. Most of the increased contribution would come from wind (2200GW, or eight times the current capacity) and photovoltaic solar (2700GW, or 28 times the current capacity!). Other renewable energies would contribute almost 1000GW of additional capacity, notably biomass (300GW) and solar thermal (400GW).

These forecasts can be looked at in light of the current LCOE as well as their projections. Both the CPS scenario and the NPS scenario seem too favorable for fossil based generation as the likely regulatory measures under preparation will lead to an increase in the relative cost of electricity for these sources. Indeed, the deployment of carbon pricing, the natural increase in the cost of capital (associated with regulatory risks), and a general increase in overnight and operation and maintenance costs linked to regulation, will all drive costs upwards. The 450 scenario and the Energy [R]evolution scenarios, assuming a decrease in coal and oil power generation capacities, seem thus more likely. The strong difference of evolution between coal and natural gas however does not seem fully realistic. Indeed, natural gas has a much lower impact on greenhouse gas emissions than coal. Coal is today the most affordable technology. More importantly, it is widely available in many countries, whereas natural gas is harder to procure. Considerations on energy independence could thus tend to not completely substitute coal by natural gas in many countries. China is a very good example of this, with over 75% of its current power generation done with coal. Now, Greenpeace assumes a complete phase out of nuclear energy by 2050, with over 80% decrease by 2030. This scenario is not realistic since nuclear power is projected to remain globally competitive over the period. The 450 scenario, which is partially based on nuclear energy, is more realistic. The consequence would be a lower ramp up of renewable energies, in particular photovoltaic solar. Now, the 450 scenario remains fairly conservative on photovoltaic solar, with "only" 600GW of additional capacity. Taking into account the strong growing competitiveness of this technology, especially when distributed in homes and buildings, the likelihood of a higher share for solar energy in the mix is fairly high.

To summarize, the 450 scenario seems closest to reality. The decline in coal might be lower than expected (closer to the forecast from the NPS scenario), and the share of photovoltaic solar higher than forecasted. Looking at the latest 5-year data from BNEF (BNEF/NEO, 2017), it appears wind and photovoltaic solar are indeed increasing fast. With the 450 scenario ranging around 10% of growth year on year till 2030, the BNEF data show a growth of around 20% on average. In 2016, wind and photovoltaic solar energies represented over 50% of total new capacity additions. However, while the 450 scenario estimates a decrease of around 1.3% per year for coal and a slight increase of natural gas of 1.7% through to 2030, BNEF has shown that coal and natural gas has kept increasing at a pace of nearly 3% per year

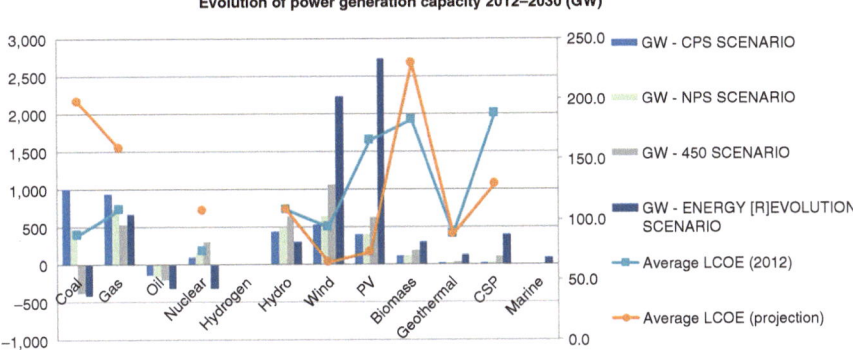

Fig. 3.36 Evolution of power generation capacity (Greenpeace 2015; © OECD/IEA, WEO 2014)

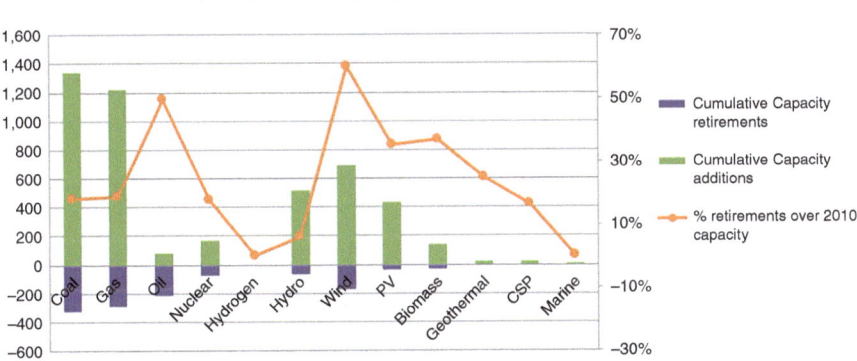

Fig. 3.37 Evolution of power generation capacity—current policy scenario (Greenpeace 2015)

since 2012, a sign that the transformation of the energy mix has not yet started in many regions around the world (Fig. 3.36).

This evolution of power generation capacity would not only meet the increase in demand. It would also be an answer to the planned refurbishments of old conventional technologies reaching end of life. Figure 3.37 shows the evolution of refurbishment and new capacity additions in the CPS scenario. Fossil based technologies reaching end of life add up to 800GW by 2030. 70GW of nuclear power plants will also be decommissioned over the period, together with 67GW of hydropower. Close to 170GW of wind farms would also reach end of life and need refurbishment. Other renewable technologies would have very low replacement rates by then. On average, around 21% of the current power generation capacity would reach end of life by 2030 and need replacement or substitution. This ratio is higher for wind farms (60%), oil power plants (50%) and photovoltaic solar (35%). It is much lower for coal, gas, and nuclear power plants (18%). The lifetime of traditional technologies such as coal, gas and nuclear is indeed around 30–40 years while that of renewable

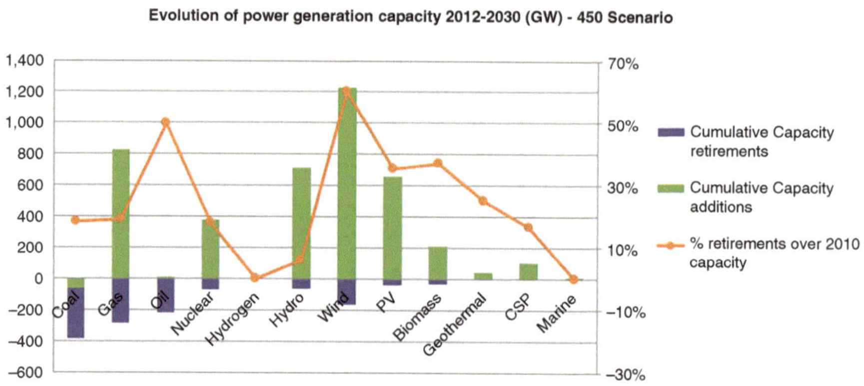

Evolution of power generation capacity 2012-2030 (GW) - 450 Scenario

Fig. 3.38 Evolution of power generation capacity—450 scenario (© OECD/IEA, WEO 2014)

energies such as wind or solar is around 20–25 years (Leonardo 2007; NREL/coal 1999; Telegraph/Wind 2016; Science Direct 2014). The retirement rate is thus higher for renewable energies, and therefore there is a need for substitution.

The CPS scenario shows that new constructions of coal power plants and natural gas power plants would largely exceed the planned refurbishments, leading to an increase in overall power generation capacity (Fig. 3.37).

The main difference in the 450 scenario is that there would not be new construction of coal power plants in the coming 20 years to compensate for the existing plants reaching end of life, leading to a net decrease of installed capacity (Fig. 3.38).

With many conventional power plants reaching progressively end of life, their likely replacement by renewable energies will lead to a massive change of the energy supply mix. Now, since the lifetime of renewable power plants is lower, the substitution rate of power generation is expected to increase significantly in the coming decades.

Further on the horizon, these changes in the power generation mix will lead to strong evolutions of the capacity factors of the different sources of power. The capacity factor corresponds to the ratio between the amount of energy supplied (TWh) and the actual generation capacity (GW). As already explained, the capacity factor is a critical indicator of the profitability of a power generation source. The more the source is able to produce power, the more the revenue. Now, some sources are more sensitive than others. Nuclear power plants and renewable energies in general are the most sensitive since the initial investment cost is more important in the corresponding LCOE. Consequently, their production output must be maximized in order to ensure a profitable investment. Conventional fossil based technologies like coal or natural gas are less sensitive as they are much more dependent on variable costs such as maintenance and fuel costs (and even carbon costs, which are evaluated on the base of the production output). The capacity factor of renewable energies and nuclear energy is generally expected to improve over the years (at worse to remain stable). An increase in the capacity factor will have a positive impact on the competitiveness of the LCOE of these technologies. Fossil based

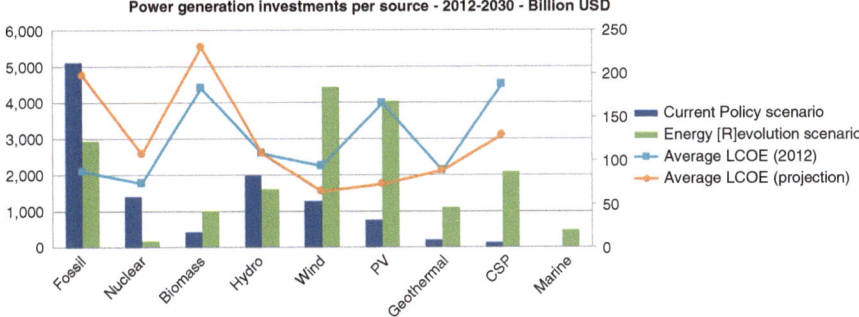

Fig. 3.39 Power generation investments by source (Greenpeace 2015; © OECD/IEA, WEIO 2014)

technologies are expected to see their capacity factor slightly decrease. Data can also be forecasted until 2050 (© OECD/IEA, WEO 2012, 2014; Greenpeace 2015). If the reference scenario from the International Energy Agency (2014) does not bring about any significant changes, the massive penetration of renewable energies fore-casted in the Energy [R]evolution scenario leads however to a further decline of the capacity factor of traditional fossil based technologies. This decline could go up as high as 30% for coal technologies and 22% for natural gas by 2050, leading to a significant impact on the profitability model of coal power, with an increased LCOE of around 10–15%. These evaluations need to be looked at regionally as capacity factors will greatly vary depending on local situations.

The corresponding level of investments required to ensure the realization of these scenarios can also be evaluated. Figure 3.39 shows the level of investments by source according to the CPS scenario and the Energy [R]evolution scenario (Greenpeace 2015; © OECD/IEA, WEIO 2014; compared to © OECD/IEA, WEO 2014). It shows that the cumulated level of investments in the CPS scenario amounts to 11.3 trillion dollars for the period 2012–2030 while the Energy [R]evolution scenario forecasts 17.9 trillion dollars. The more "offensive" the scenario on renew-able energies, the more expensive the investment. In the Energy [R]evolution scenario, the amount of investments on renewable energies represents three times the amount in the CPS scenario. The low capacity factor of renewable energies indeed requires a higher volume of GW installed for the same energy supplied. The natural consequence is a more important level of initial investment. In both scenarios a significant amount of investments remain on traditional power generation units (fossil). The Energy [R]evolution scenario as well assumes a close to complete phase out of nuclear. This assumption could be challenged and investments in nuclear could turn out to be more significant, in particular in certain regions of the world (OECD Asia, Europe, etc.).

3.2.2   Regional Perspectives on Power Generation Investments

The dynamics of power generation construction and investment will vary from one region to another (Fig. 3.40). Generally, some patterns can be identified, depending on the scenario considered. All scenarios from the International Energy Agency (2014) present only slight differences of the overall power generation capacity construction in each region, with the exception of China. In China, the CPS scenario is much higher than the 450 scenario, with 300GW additional capacity. Greenpeace presents a very different scenario, with much more power generation construction. It considers notably a power construction market in North America five times bigger than the scenarios from the International Energy Agency (2014). Apart from North America, most forecasts from Greenpeace (2015) range between two and three times the actual ones from the International Energy Agency (2014). In particular, India represents a massive difference in investments, with over 400GW of additional capacity between the scenarios. Only Europe remains fairly stable across scenarios, with a Greenpeace forecast not far from the ones of the International Energy Agency (2014). In the end, the massive increase in power generation capacities will be in China (up to 1400GW), which would double its capacity, and in India (between 400GW and 800GW), which would multiply it between two and four times. North America could also bear massive investments. Greenpeace estimates the increase in power generation capacity could reach 71% of the current 2012 production, compared to a 19% increase from existing forecasts from the International Energy Agency (2014).

This evolution of the power generation capacity also follows the need to replace old power plants reaching end of life. It appears clearly on Figs. 3.41 and 3.42 that the regions with the most important needs in terms of replacement are OECD countries, in particular North America and Europe, which are pushing naturally

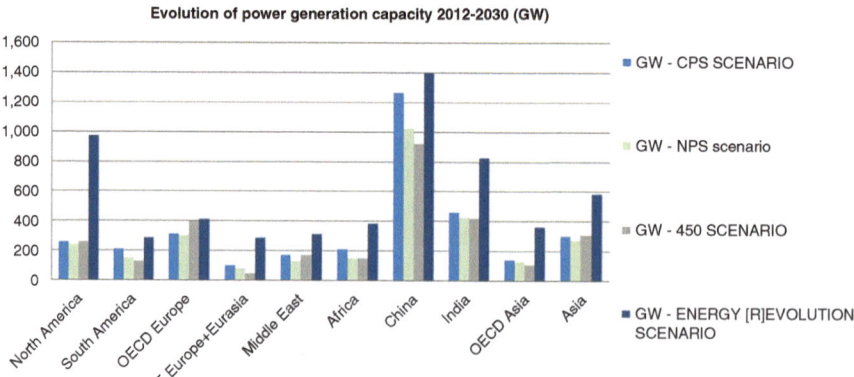

Fig. 3.40 Evolution of power generation capacity by region/country (Greenpeace 2015; © OECD/ IEA, WEO 2014)

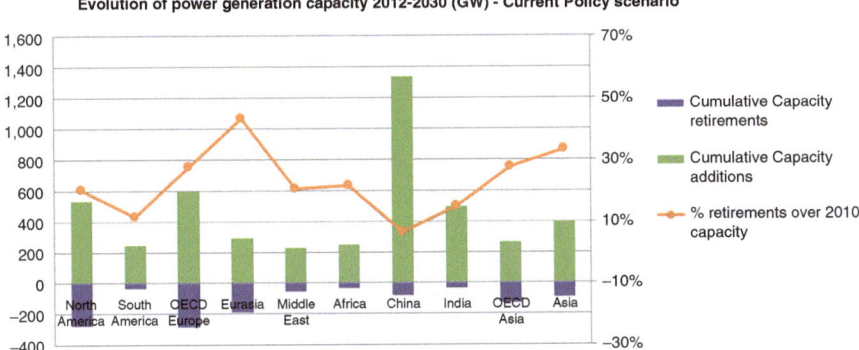

Fig. 3.41 Evolution of power generation capacity—current policy scenario (Greenpeace 2015)

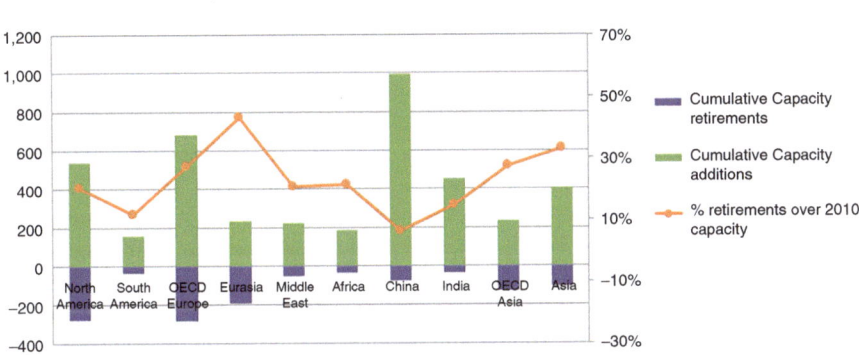

Fig. 3.42 Evolution of power generation capacity—450 scenario (© OECD/IEA, WEO 2014)

the construction market up, as well as accelerating the change of mix, taking into account the new competitive paradigm of power generation. In percentage terms, Eurasia and Asia (including OECD Asia) are the regions with the greatest needs in terms of power generation replacements, which could then lead to the most important change of mix. In summary, OECD countries have a fairly high rate of existing power plant retirements, which will lead to a change of mix of power generation as new installations could use renewable sources. In non-OECD countries, with the exception of Asia and Eurasia, where the rates of retirements are high, most of the installations will be to meet the strong increase in demand. The change of mix in power generation could thus occur from the actual mix of the new power generation construction, without substitution from old sources.

Investment levels follow power generation construction trends. Overall, forecasts estimate an increase in power generation investments of around 56% in the CPS scenario (© OECD/IEA, WEIO 2014), and up to 96% in the robust scenario from

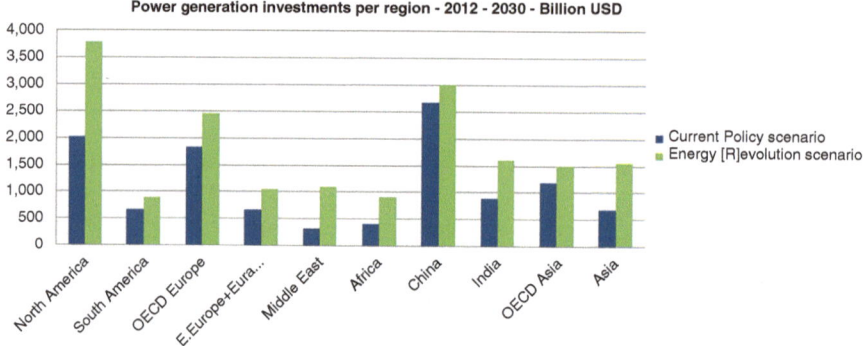

Fig. 3.43 Power generation investments by region/country (Greenpeace 2015; © OECD/IEA, WEIO 2014)

Greenpeace (2015) (Fig. 3.43). The lowest increase in activity will be in Europe, with an increase in investments of between 9% (CPS scenario) and 25% (Energy [R] evolution scenario). This is due to the fact that investments in Europe in the previous period (2000–2012) were among the highest on the planet, with over 980 billion dollars invested, almost as much as China. The baseline is therefore extremely strong. Investments in China will follow a similar path, with growth estimated around 55% in all scenarios. Investments in China should remain the highest on the planet for the coming period. In other regions, the acceleration in investments is faster because lower investments were made in the previous period. This is the case for other OECD regions such as North America (between 28% and 117%) and OECD Asia (up to 94%). Investment growth in India is expected to range between 133% and 164% compared to the previous period to meet the increase in demand, which is expected to be multiplied by four in 20 years. In the Middle East, investments are expected to grow between 69% and 264% in the Energy [R] evolution scenario! In Asia, investments would more than double in all scenarios. In Eurasia, investments would be close to double in all scenarios (between 86% and 99%). Finally, investments in Africa are expected to skyrocket and multiply by close to four times.

China represents the region with the highest volume of investments in the CPS scenario (17–24% of total investments). It will be followed by North America, where investments are expected to increase significantly its (18–21%), and Europe (14–16%). There is strong uncertainty regarding the forecast of investments for North America since the cost of the transition to renewables (Energy [R]evolution) represents for the region a very important gap in terms of volume of investments compared to the "business as usual" scenario.

3.3 High Penetration of Renewable Electricity

3.3.1 The Renewable Electricity Mix

The massive investments in renewable energies have led to an increase in their share in the overall power generation mix.

Renewable energies accounted for 21% of the total energy (including hydropower) produced in 2012. The region with the highest adoption level is South America, with already 60% of the power consumed produced by renewable energies, thanks to its large hydropower base. Europe comes second, with nearly 30% of its electrical energy produced by renewable energies. The mix in Europe is more balanced, with photovoltaic solar and wind farms accounting for around 8% together. Other countries range below 20% today. Hydropower is an essential contributor in terms of energy produced. Wind is deployed in North America, China, India and OECD Asia where it accounts for between 1% and 2% of the electrical energy consumed in average. Photovoltaic solar has a smaller role, with around 1% of the total contribution in terms of electrical energy consumed in the world. Other renewable sources such as biomass, solar thermal (CSP), or marine account for 2% on average, except in Europe, where they account for as high as 5% of total electrical energy consumption. Nevertheless, it is worth repeating here that wind and photovoltaic solar energies represented over 50% of new capacity additions in 2016 (BNEF 2017), a clear change of paradigm.

The overall contribution of renewable energies in the electrical energy consumption mix is expected to increase in all scenarios presented (Fig. 3.44). It would grow from 21% on average to 25% in the CPS scenario, 42% in the 450 scenario, and up to 58% in the Energy [R]evolution scenario (© OECD/IEA, WEO 2014; Greenpeace 2015). The rate of penetration of renewable energies thus continues to increase. The scenario from Greenpeace (2015) is however more disruptive, with an increase in the

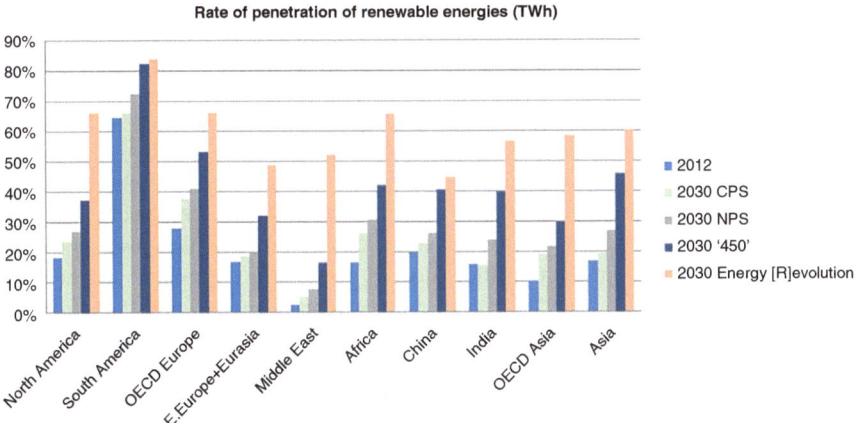

Fig. 3.44 Rate of penetration of renewable energies (Greenpeace 2015)

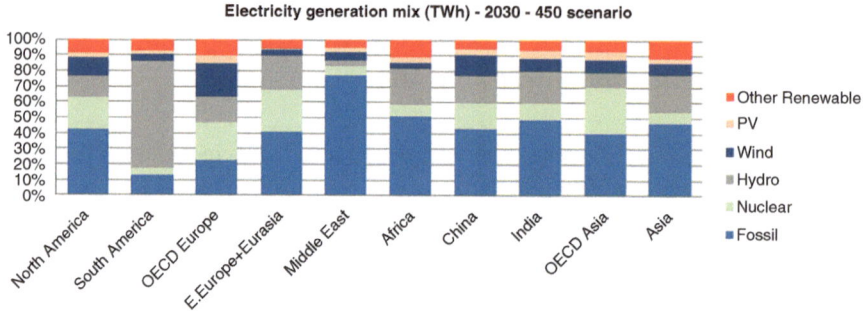

Fig. 3.45 Electricity generation mix in 2030–450 scenario (© OECD/IEA, WEO 2014)

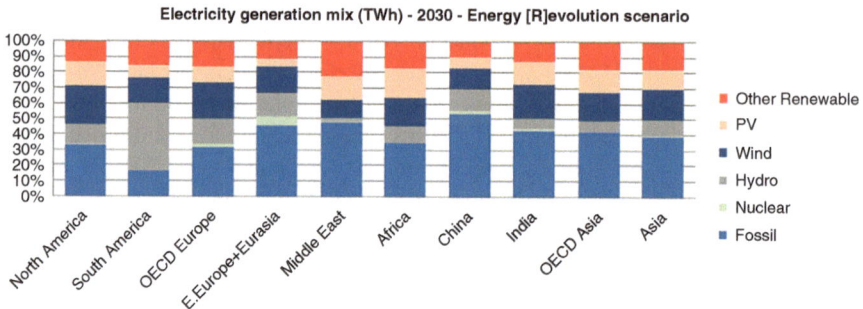

Fig. 3.46 Electricity generation mix in 2030—Energy [R]evolution scenario (Greenpeace 2015)

rate of penetration of renewable of up to 58%, compared to 21% in 2012. In this scenario, all countries increase significantly their mix of renewable energies.

Most of this growth in the Energy [R]evolution scenario can be explained by the accelerated penetration of wind and photovoltaic solar into the grid. While their combined share does not exceed 14% on average in 2030 according to the 450 scenario, compared to 3% today, it would skyrocket up to 30% in the scenario presented by Greenpeace (2015) (Figs. 3.45 and 3.46). This corresponds to around 10,000 TWh of wind and solar by 2030, against 600 TWh today worldwide, and a forecasted 4300 TWh in the 450 scenario. The scenario from Greenpeace (2015) thus seems extremely ambitious. It is indeed founded on the assumption that both coal and nuclear power could be completely phased out by 2050. By then, renewable energies overall would represent 92% of the total electricity generation. This also represents an additional investment of 6 trillion dollars worldwide compared to the CPS scenario and an additional 2 trillion dollars compared to the 450 scenario. Greenpeace (2015) thus suggests that the 450 scenario could be transformed into the Energy [R]evolution scenario for an additional 30% investment and modify completely the electricity generation mix. With current forecasts on solar and wind costs, this scenario is becoming more likely over time.

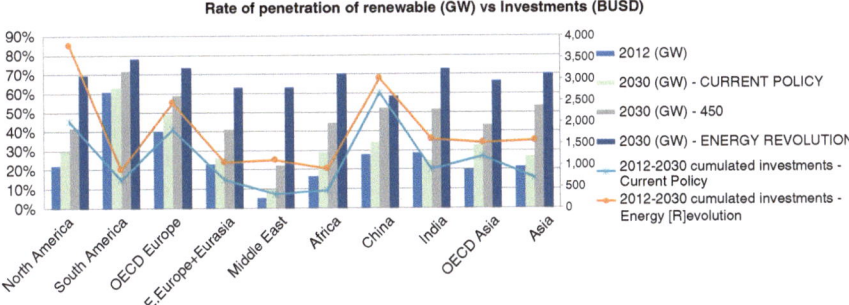

Fig. 3.47 Rate of penetration of renewable vs. investments (Greenpeace 2015; © OECD/IEA, WEIO 2014; © OECD/IEA, WEO 2014)

This analysis of the penetration of renewable energy uses the actual volumes of energy (TWh) produced by different sources of power to calculate the overall mix. This gives an indication of how electricity is supplied, and is useful when it comes to evaluating the average price of electricity traded, as well as the relative share of the power sector in greenhouse gas emissions. However, it does not reflect well the volumes of investments required across sources, as this can only be evaluated based on the actual capacities (GW) installed. The global evolution of renewable capacity is quite different from that of energy. In 2012, renewable represented 21% of all energy consumed and 28% of the total power generation capacity installed. Forecasts differ significantly. The CPS scenario estimates that up to 25% of the energy consumed in 2030 could be from renewable sources. This would correspond to up to 34% of the power generation capacity installed. In the 450 scenario, the 42% attributed to renewable energy would correspond to 49% of the power generation capacity installed. In the Energy [R]evolution scenario, the respective percentages would be 58% and 68% (Fig. 3.47). In the end, the divergence between the rate of penetration in energy and capacity terms can be explained by the low capacity factor of renewable energies such as wind or photovoltaic solar. That is, with renewable sources, more capacity is needed to produce the same amount of energy than with conventional sources. Now, this evolution of the power generation capacities will strongly depend upon the level of investments in a given region. Even though the "high renewable" scenario from Greenpeace (2015) shows higher investments, the gap between this scenario and the CPS one varies from one region to another. The jump to a high renewable mix is particularly high in North America and the Middle East. In North America, moving to a high share of renewable represents almost a doubling of the level of investments.

Finally, variable renewable energies (wind and solar) do not operate continuously on the network. It is extremely difficult to evaluate the total amount of renewable energies that could be produced at the same time on the grid as this depends on a number of factors, among them weather conditions, operation and maintenance schedules, as well as actual market dynamics. An overly high penetration rate at a given time leads other plants to shut down unexpectedly or to reduce their

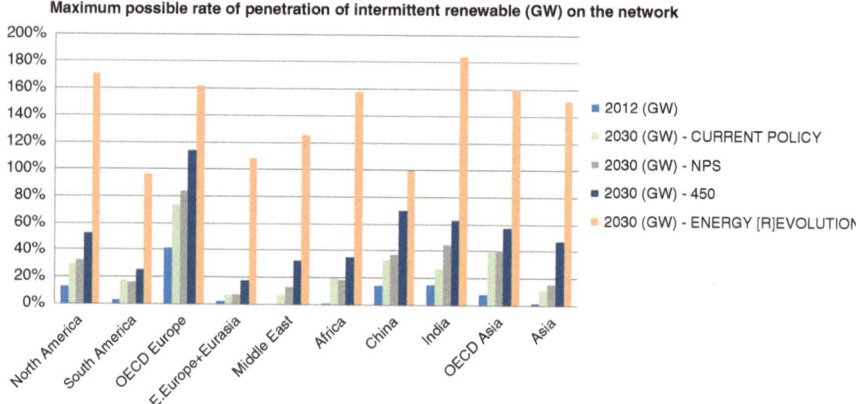

Fig. 3.48 Maximum possible rate of penetration of intermittent renewable (Greenpeace 2015; © OECD/IEA, WEO 2014)

production output. A too low penetration rate forces "peak-load" power plants (very expensive and polluting) to be started in order to meet the net demand. The unpredictable production of variable renewable sources can thus have a massive impact on the way the market operates, and beyond actual real-time grid balancing. We estimate here the maximum theoretical rate of penetration of variable renewable energies by dividing the sum of the capacities of wind and solar (assuming they operate together at their maximum power at a given time of the day) by the actual average net capacity demand, which can be obtained by converting the overall yearly energy consumption (TWh) to an average capacity requirement (GW). The figure is obviously highly theoretical as on one hand renewable energies never operate together, and on the other hand the load continuously varies during the day. Nevertheless, the results show a significant change of paradigm on how to operate grid networks (Fig. 3.48). The 2012 worldwide average is 15% of penetration, this means that no more than 15% of the power consumed by the network can be supplied by variable renewable energies at any given time of the day and given any operating condition. Now, this ratio is significantly higher in Europe, the only region of the world with a significant share of both wind and photovoltaic solar power. If all these capacities were operate at their maximum output at the same time, 41% of the power demand could be supplied by these variable sources at a given time. However, this would present load balancing issues in certain parts of Europe. Now, this ratio is expected to skyrocket in all regions of the world in the coming years. The CPS scenario, a low base scenario of renewable penetration, shows that North America, China and OECD Asia would meet similar issues in the coming years, as their rate would jump to 29% for North America, 33% for China and 39% for OECD Asia, respectively. In this conservative scenario, the rate in Europe would increase to 74%, leading to even more stringent issues. The 450 scenario plans for a more massive deployment of renewable energies throughout the world. Many regions such as the Middle East, or Africa would increase around 30%. Asia and

India would increase significantly, between 47% (Asia) and 63% (India). Only South America (23%) and Eurasia (16%) would keep a rather low volume of variable renewable energies. Finally, OECD countries and China would increase significantly their rate of penetration. For Europe, this rate could be higher than 100% of the total demand! Finally, the scenario from Greenpeace (2015) evaluates a high share of variable renewable in the overall capacity mix, leading to very high ratios—well above 100% for almost all regions of the world—and leading to a clear change of paradigm on how grids are managed.

To summarize, the rate of penetration of renewable can be evaluated in different ways (Table 3.1). It can be estimated from the actual energy produced (TWh). In such a case, it is an interesting indicator of the balance of electricity prices as well as of the overall level of greenhouse gas emissions from the power sector. It can also be evaluated from the actual capacities installed or to be installed (GW). It then becomes an interesting indicator of how investments are being made and the overall level of the investments needed. Finally, a more sophisticated evaluation involves computing the maximum possible share of production of variable renewable energies (the ones that cannot be accurately forecasted) at a given time of the day (Max VRE). The inherent intermittency of these energies yields a number of grid management issues as well as more structural evolutions of the power generation mix (Fig. 3.49). These critical topics at the heart of the transition in progress will be reviewed in the Chap. 4.

3.3.2 The Dawn of Distributed Generation

Distributed generation defines the power that can be generated from renewable energy sources in residential homes, commercial buildings and light industries. It mainly refers to photovoltaic solar technology installed on the rooftops of houses and buildings. Thanks to various subsidies, early adopters have in the past installed rooftop solar systems in order to produce part or all of their electricity consumption. The situation has been changing in the past few years. Indeed, the costs of solar modules have been divided by five in the 2008–2013 period (© OECD/IEA, Solar 2014). The system costs of solar have also been divided by three, but they still vary significantly from one region to another, because of the constraints of local regulations. Generally, this past evolution has modified significantly the situation with regards to the installation of rooftop systems. The LCOE of rooftop systems could be retrieved in certain regions such as North America, Europe and OECD Asia and mapped to the retail prices in various countries (ANEEL 2016; BOI 2016; Brazil Business 2016; Business Tech 2016; DEWA 2016; EIA Price 2016; EWB 2016; Global Climate Scope 2016; Hong Kong Electric 2016; © OECD/IEA, Costs 2015; Kepco 2016; Lumo Energy 2016; Manitoba Hydro 2016; MB 2016; Ovo Energy 2016; PLN 2016; Saudi Electric 2016; Shenzhen 2016; Singapore Power 2016; Storm Report 2016; Sudan Tribune 2016; Times of Israel 2016; TNB 2016; TurkStat 2016; Vietnam News 2016). These retail prices can be averaged in order to provide a

Table 3.1 Rate of penetration of renewables

Country	2012			2030 CPS			2030 '450'			2030 energy [R]evolution		
	TWh %	GW %	Max VRE %	TWh %	GW %	Max VRE %	TWh %	GW %	Max VRE %	TWh %	GW %	Max VRE %
North America	18%	22%	13%	24%	30%	29%	37%	42%	53%	66%	70%	171%
South America	65%	61%	3%	66%	63%	18%	82%	72%	25%	84%	78%	97%
OECD Europe	28%	40%	41%	38%	49%	74%	53%	59%	114%	66%	73%	162%
E.Europe +Eurasia	17%	23%	3%	19%	26%	7%	32%	41%	18%	49%	63%	108%
Middle East	2%	5%	0%	5%	10%	7%	17%	22%	32%	52%	63%	126%
Africa	16%	17%	1%	26%	29%	20%	42%	44%	35%	66%	70%	158%
China	20%	28%	14%	23%	34%	33%	40%	52%	70%	44%	59%	100%
India	16%	29%	15%	15%	25%	26%	40%	52%	63%	56%	73%	184%
OECD Asia	10%	20%	8%	19%	33%	39%	30%	43%	57%	58%	57%	160%
Asia	17%	21%	1%	19%	27%	11%	46%	54%	47%	60%	70%	151%
Total	21%	28%	15%	25%	34%	31%	42%	49%	60%	58%	68%	139%

Source: Author's own calculation based on data from IEA and Greenpeace

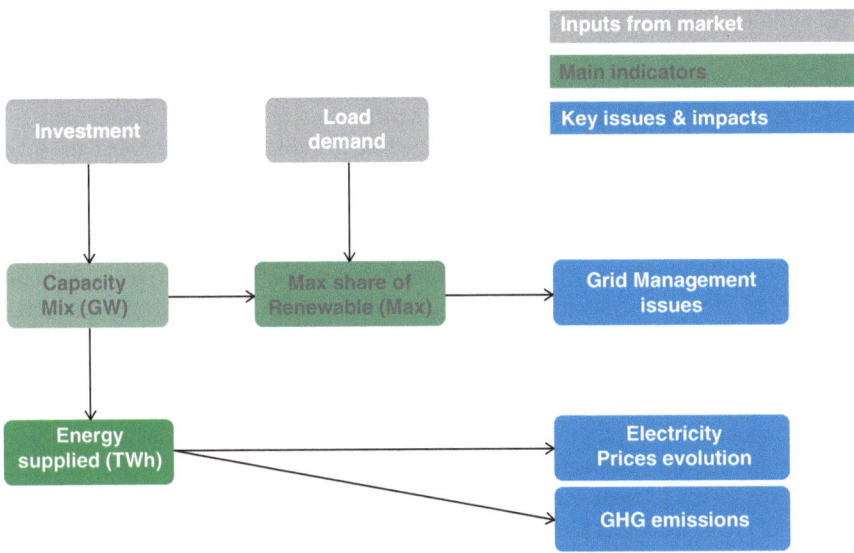

Fig. 3.49 Renewable penetration in the grid

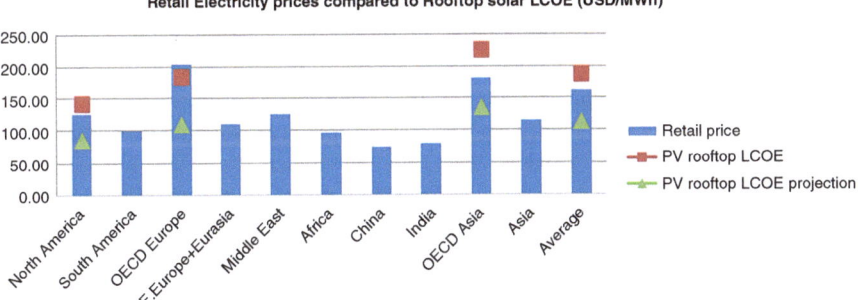

Fig. 3.50 Retail electricity prices compared to rooftop solar LCOE (AER 2016; ANEEL 2016; BOI 2016; Brazil Business 2016; Business Tech 2016; DEWA 2016; EIA Price 2016; European Commission/Wholesale 2016; EWB 2016; Global Climate Scope 2016; Hong Kong Electric 2016; JPEX 2016; © OECD/IEA, Costs 2015; Kepco 2016; Lumo Energy 2016; Manitoba Hydro 2016; MB 2016; Ovo Energy 2016; PLN 2016; Saudi Electric 2016; Shenzhen 2016; Singapore Power 2016; Storm Report 2016; Sudan Tribune 2016; Times of Israel 2016; TNB 2016; TurkStat 2016; Vietnam News 2016)

simplified picture by region. Obviously, they strongly vary from one country to another. Typically, they vary in South America from 100USD/MWh in Brazil to 146USD/MWh in Colombia. They also vary significantly in Europe from 132USD/MWh in Hungary to 360USD/MWh in Denmark. They also vary in Asia between Malaysia (63USD/MWh) and the Philippines (179USD/MWh). The prices are very low in Saudi Arabia (no more than 23USD/MWh) and reach ten times this value (230USD/MWh) in United Arab Emirates. Despite those differences, averages are provided to simplify the picture. A global average is also computed (Fig. 3.50).

This simplified view generally shows that if the LCOE is still higher today in North America and OECD Asia on average, it is already lower than the retail prices in most countries of Europe. This is leading to a natural development of these solutions across residential homes and buildings. When looking at the LCOE projections for rooftop solar, taking into account the forecasted evolution of the cost of solar modules as well as the system costs, it becomes clear that rooftop solar technologies will become competitive everywhere with regards to the current retail prices, leading to broader development of this technology.

Distributed generation is thus slowly emerging as a critical alternative to traditional retail electricity purchased from electricity retailers, reducing by the same amount the actual wholesale market of electricity generation. This is another evolution which is both disrupting traditional electricity markets' structure as well as the typical power distribution architecture.

The potential of deployment of distributed generation can be evaluated. The exercise first requires evaluating the actual surface area of residential and building areas by region. Sources exist which help evaluate the actual surface area of residential areas per capita by region. The worldwide average in 1970 ranged around $14m^2$/capita and has kept increasing since. Regions such as North America benefit from higher surface area per capita, around $80m^2$/capita (California Photon 2010). Other regions range below North America, with strong variations between countries, new economies being usually smaller at around $25m^2$/capita while mature economies typically range around $36m^2$/capita (HOTG 2006). Overall, it can be estimated that the worldwide residential surface area averages $30m^2$/capita today. From this, the actual rooftop area can be deducted by applying a coefficient to it, typically dividing by two to three. A coefficient of three is typically applied in new economies and a coefficient of two in mature economies to match the actual rooftop areas evaluated. A coefficient of 1.5 is applied in North America (California Photon 2010). It is then considered that the rooftop surface of commercial buildings is similar to that of a residential one, according to a Navigant study done in the United States (California Photon 2010). These assumptions help evaluate the overall rooftop surface area available in each region. From this, an average coverage of this surface is applied, which differ between residential areas (22% was considered for all regions) and commercial buildings' areas (65% was considered for all regions). This leads to an overall potential surface area for rooftop solar systems. The actual yearly "production" of this potential can then be estimated by multiplying the solar flux (which varies from one region to another) to the actual conversion efficiency of solar modules, considered to be 25% on average worldwide (California Photon 2010). Solar flux can vary significantly from one region to another, as well as across seasons of the year (PVEducation 2016). Worldwide, it averages $1200kWh/m^2$/year in winter, with lows of $730kWh/m^2$/year in North America or OECD Asia, and $182kWh/m^2$/year in Eurasia. In summer, it will range around $1600kWh/m^2$/year, with a peak in the Middle East ($2500kWh/m^2$/year) and a low base in the south hemisphere (e.g., $730kWh/m^2$/year in South America). This data helps compute the overall potential amount of energy (TWh) that could be produced by distributed rooftop solar systems. Again, it is considered that this volume of energy is equally

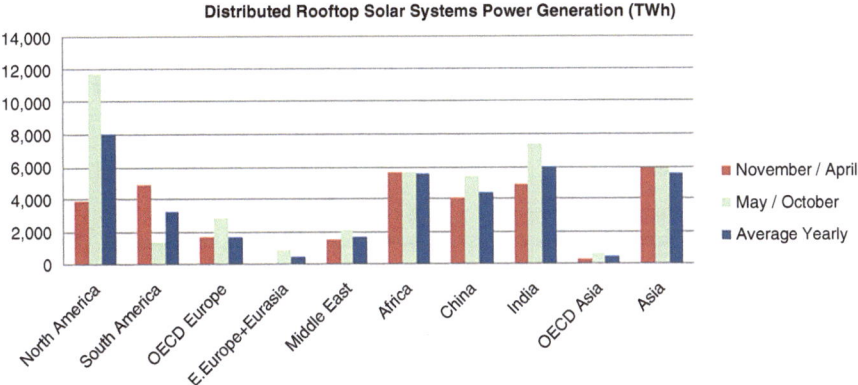

Fig. 3.51 Distributed rooftop solar systems power generation (California Photon 2010; PVEducation 2016)

split between residential homes and commercial buildings. It shows a potential of around 37,000TWh of electricity generation, with a low 32,000TWh in winter and a high 44,000TWh in summer (Fig. 3.51). This figure is to be compared to the current 17,000TWh of worldwide electricity consumption in the world, and the forecasted 33,000TWh by 2035 (CPS). Distributed generation thus represents a fantastic potential.

Currently, North America is leading in terms of power generation capacity essentially because it has the highest rooftop surface area per capita in the world. As economic development continues and new economies see an increase in their living standards, it is very likely that the average rooftop surface area available per capita will also increase, leading to a fantastic increase in the potential for further distributed generation.

The ratio of rooftop solar systems' electricity generation potential to actual electricity consumption can then be retrieved to define to which extent a region is exposed to the development of rooftop solar systems. If above one, it means there is more potential of electricity generation from rooftop solar systems than there is actual electricity consumption. This ratio was calculated for 2010, as well as for 2035, using the forecasted electricity consumption from the CPS scenario of the International Energy Agency (2014) (Fig. 3.52). The ratio of Africa and India are way above ten, and the figure has been clipped. The figure shows that distributed solar generation could potentially meet the actual electricity consumption demand in most regions of the world, with the exception of Europe, Eurasia, China and OECD Asia (excluding Australia). It thus could not technically present a significant disruption of the market structure in these last regions.

The forecasts of realization of this potential can then be evaluated. Roland Berger (2015) has estimated the share of rooftop solar systems in the overall solar installations. This share varies across regions. It gets as high as 72% in Europe, against 16% in Middle East and Africa, two regions privileging large utility-scale farms. It breaks even in China, gets as high as 59% in Asia Pacific and is lower at 39% in the

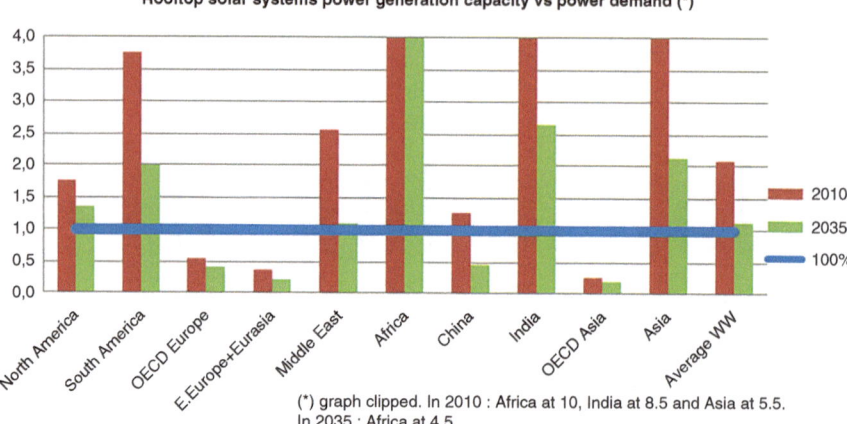

Fig. 3.52 Rooftop solar systems' capacity vs. demand (California Photon 2010; PVEducation 2016; Greenpeace 2015)

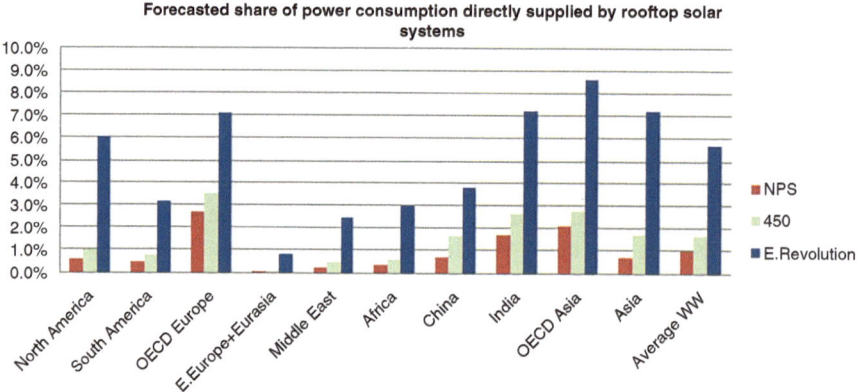

Fig. 3.53 Forecasted share of electricity consumption directly supplied by rooftop solar systems (Greenpeace 2015; Roland Berger 2015)

Americas. The actual share of rooftop solar systems' electricity generation can then be estimated applying this ratio to the actual forecasts of solar electricity generation in the International Energy Agency (2014) and Greenpeace (2015) scenarios. Figure 3.53 shows the actual share of rooftop solar systems' electricity as a percentage of total electricity consumption. This share does not exceed 2% in all scenarios from the International Energy Agency (2014), but ranges above 5% in most regions of the world in the Energy [R]evolution scenario. It gets as high as 8% in OECD Asia and 7% in Europe and India.

This can be compared to the data available from other sources. In North America, Chadbourne (2014) estimates this potential could reach 7% of the total electricity consumption, with some utilities facing a much greater impact, such as Arizona

Public Service (34%), Pacific Gas & Electric (26%) or San Diego Gas & Electric (25%). Other scenarios foresee solar energy being as high as 1,000TWh between 2035 and 2050 in the United States (Deloitte 2015; NREL/Solar 2015). This would correspond to 16% of the total electricity generation by 2035. Assuming a 60/40 split between large utility-scale and rooftop solar systems (Roland Berger 2015), this yields around 7% of electricity produced by rooftop solar systems on average. The forecasts from Greenpeace (2015) range around 6%. In Europe, Roland Berger (2015) estimates the share of distributed generation could be as high as 9% (compared to 7% in the "Energy [R]evolution" scenario). The share in Australia could be as high as 10% of the total power consumed by 2030 (Parkinson 2015; AEMO 2012), backing up the average estimation of 8% for the whole OECD Asia computed from the Greenpeace (2015) projections.

McKinsey (2012) estimates that the forecasted additional generation of rooftop solar systems could be as high as 600GW by 2020, one of the most aggressive forecasts. For the United States only, McKinsey expects 130GW of rooftop solar systems to be installed by 2020, while Credit Suisse (2015) estimates this figure could be almost twice higher at around 225GW. In terms of energy consumption (TWh), this would correspond, assuming capacity factors of around 11% for the current period, to an average 1.7% of total electricity consumption worldwide. This ratio would reach 2% in North America and 4% in Europe and Africa.

In summary, distributed generation offers considerable potential. In many regions of the world, the potential for self-generation could top the 2035 forecasted overall consumption of electricity. Theoretically, most of the electricity consumption could thus be substituted by local distributed generation, with the exception of a few regions such as Europe, China or OECD Asia. This is another disruption that the deployment of competitive variable renewable energies introduces to traditional electricity markets. Self-generation would indeed yield significant load defection from and thus revenue drops to traditional operators, leading to important market restructuring. In addition, local distributed generation offers the considerable advantage of requiring less from the transmission and distribution infrastructure. It is thus theoretically much cheaper to procure, and thus potentially a fantastic source of productivity, provided the grid infrastructure is never required to supply the full amount of load at any time of the day (see Chap. 4). In practice, this is where grid parity is likely to first be reached in many countries. Over time, the development of distributed generation could thus lead to the redesign of transmission and distribution infrastructures. These would have to evolve towards an infrastructure that interconnects islands of self-generation. These islands (also called microgrids) would operate autonomously, connecting to the grid to exchange power capacity in both directions, depending on the balance of the overall grid. In many regions of the world (South East Asia, Africa, Australia, South America, the United States), this has already proven to be an economic alternative to traditional grid expansion, and proves that solutions exist (see Chap. 4) to cope with both the intermittency of renewable production and the inelasticity of demand. While the potential development of distributed generation is considerable, current forecasts are much more conservative. The share of electricity that could be directly supplied by distributed generation is not expected to exceed 6% in the coming years, with significant differences across countries and utilities. Some utilities could face a much higher share of substitution while others would virtually not be impacted. The development of distributed generation, already competitive in many regions of the world, will however continue to intensify in the coming years, leading to global competitiveness of this model of energy production. Provided regulations enable its deployment, the forecasts could be proven wrong and the untapped potential of this clean source of power realized to a much wider scale.

3.4   Summary: Cost of Transiting to a New Power Mix

The key indicator of electricity generation markets is the LCOE, which can be calculated for each source of power in each region or country. The LCOE enables the comparison of various sources of power and their sensitivity to a number of factors. It does not necessarily reflect current electricity prices, as the actual existing power generation fleet in a given country can already be amortized. In such cases, electricity prices would tend to be lower as they would not have to account for investment costs anymore. However, it clearly reflects the required price for a given new investment to be profitable. The LCOE is made up of a number of components. The cost of electricity of a given power generation source is dependent on investments' costs (also called overnight costs), the cost of the capital, operation and maintenance costs, fuel costs (when required), as well as the capacity factor (number of hours of operation per year). It is also sensitive to various taxes and regulatory measures such as carbon pricing.

Traditionally, conventional fossil fuel based technologies (coal, gas, oil) have had low costs of electricity and therefore have been widely deployed across the world. Nuclear power is also interesting, with low costs of electricity (although usually nuclear plant decommissioning costs are excluded). Renewable energies have not been widely deployed across the world, with the exception of hydropower, which has been competitive for decades. New renewable energies have emerged over the past few years. Wind technologies have followed a learning curve and are now ranging around similar or even lower costs of electricity than conventional technologies. More recently, photovoltaic solar technologies have started to progress and the learning curve as well as the scale effect have led to a spectacular decrease of the cost of such technologies. In many regions of the world, they are just a few years away from ranging among the cheapest sources of energy (BNEF 2017). Other renewable sources such as geothermal, biomass, concentrated solar and marine are also interesting. However, with the exception of geothermal, they are often not competitive.

These sources show a different sensitivity to the various factors which influence the cost of electricity. Nuclear power and renewable technologies are generally extremely sensitive to overnight costs, costs of capital and capacity factors. Once the investments are realized, they must operate as continuously as possible, beyond the capacity factor estimated in the payback calculation. Their variable costs are relatively low in comparison to upfront costs and maintenance fixed costs. Conventional technologies are more flexible and less dependent on the capacity factor. A greater share of the cost of electricity is devoted to variable costs such as operation, maintenance and, more importantly, fuel costs. The unpredictable evolution of fuel costs has a massive impact on the profitability of those plants. They however can more easily adjust their capacity factor, and are therefore more flexible to operate on the grid.

The costs of electricity are expected to evolve in the coming decades. Typically, photovoltaic solar costs, which are not down the learning curve yet, are expected to drop by another 40–60% in the coming decades; the more recent evolution even suggests these are too conservative. Wind costs are also expected to drop by around 20%. At the same time, the costs of conventional technologies are expected to increase, essentially because of the uncertainty of their future and actual capacity factor, leading to increased costs of capital, as well as regulatory measures which could penalize their greenhouse gas emissions. Depending on the scenarios, they are expected to increase between 10% and 40%. These evolutions will deeply change the competitive landscape of electricity markets, leading to a much greater share for renewable energies, in particular photovoltaic solar and wind.

Additionally, the LCOEs evaluated for new power plants remain higher than current wholesale prices everywhere. This is because existing generation units are for the most part amortized in many regions of the world, leading market prices towards the marginal cost of production of these units. In OECD countries (in particular, Europe), the general excess of capacity over the grid adds up to this issue. Wholesale prices, as they are, will either have to increase or the increase be compensated with other subsidies (taxes for instance) or market structures (capacity markets) in order to finance the transition. This is less of an issue in new economies, where demand lifts up prices closer to the LCOE of new technologies, and overall this trend could be mitigated by the significant evolution of costs of renewable technologies, particularly photovoltaic solar.

The actual competitive landscape must though be approached with caution as it varies across regions, countries, and often within a country itself. For instance, North America, where two thirds of generated electricity comes from fossil fuels, with already fairly balanced costs of electricity, is expected to see a growing share of renewable energies in its power mix. Natural gas or nuclear power would however remain competitive. Europe, where traditionally the mix has been more balanced between technologies, shall experience an acceleration of the penetration of wind and photovoltaic solar in the mix. A lot will however depend on the actual policy around nuclear power in the region. China depends for more than 75% of its electricity needs on fossil fuels. Coal has been traditionally the main source of power in the country. However, the costs of electricity do not reflect this reality. Already, hydropower, wind, photovoltaic solar and nuclear are on par with conventional technologies. Were regulatory measures lead to an increase in the actual cost of electricity of conventional technologies, the competitiveness of these sources would be reinforced. Given the massive need for power in the country (China should multiply its consumption by more than two in the coming decades), the mix there will likely evolve towards a much greater share for renewable energies (wind, photovoltaic solar) and nuclear power. Finally, OECD Asia is dependent for 75% of its electricity consumption on fossil fuels (coal, natural gas and oil). Renewable energies seem more expensive to generate in the region, and nuclear power remains competitive. The mix there should also likely evolve towards a greater share for renewable with a strong increase in nuclear power, provided regulatory authorities do not rule against it.

In the end, renewable energies are expected to take a much more important share of electricity generation markets in the coming years almost everywhere. The competitiveness of wind and, most importantly, photovoltaic solar is often already there and in all cases only a matter of time. Various forecasts show that, in terms of energy consumed (TWh), renewable energies' penetration ratio could increase from 21% today (including hydropower) to 25% (CPS scenario), 29% (450 scenario), and up to 58% (Energy [R]evolution scenario). This ratio is slightly different when considering the actual power generation capacities installed (GW). In that case, it would go from 28% today to 34% (CPS scenario), 49% (450 scenario), or up to 68% (Energy [R]evolution scenario). Nuclear is also expected to keep a significant role, in particular in non-OECD countries (but not only!), where power consumption costs will increase significantly. Its cost remains indeed competitive. Were renewable energies costs to drop further, this could undermine the further development of nuclear energy in the coming decades. These scenarios will now be more or less achievable, depending on the actual capacity of regions to unlock the necessary investments.

The level of investments in power generation required over the period is expected to rise from around 480 billion dollars a year in the first decade of the twenty-first century to between 750 and 950 billion dollars a year in the coming 20 years, depending on the scenario (Fig. 3.54). This would be a spectacular increase in the investments in the sector, mostly driven by the replacement of retiring capacities as well as the introduction of renewable energy capacities, which have a higher ratio of

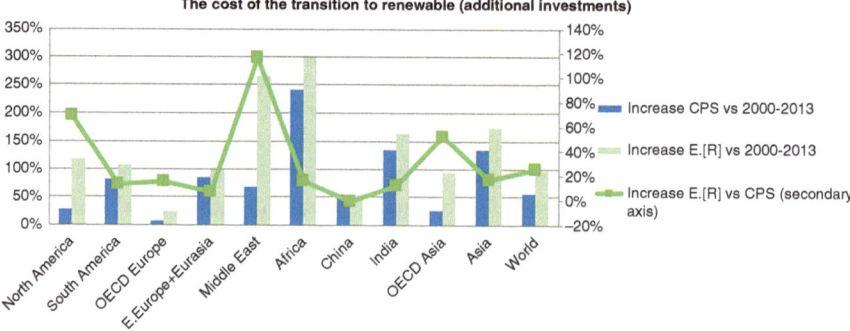

Fig. 3.54 The cost of transition to renewable (Source: author's own calculation based on data from IEA and Greenpeace)

investment overall, considering their low capacity factor. Non-OECD countries are the most impacted by this increase, in particular Africa, India and Asia, essentially because of the pressing need for energy access. The cost of transition to a high mix of renewable can be estimated by looking at the difference between the high-base scenario (Energy [R]evolution) and the low-base scenario (CPS). It would amount to 200 billion dollars a year of additional investments, or around 6 trillion dollars over the 2010–2035 period. In several regions of the world, such as Europe, South America, Africa, China, and India, the additional investment required would not represent a significant portion of the investments already locked in. This means that such a transition is extremely likely. In other regions, however, this additional investment could be prohibitive. This could be the case in North America, with an additional 70% increase in investments over the period, the Middle East or, to a certain extent, OECD Asia.

Going further, part of these renewable energies are expected to be distributed. Distributed generation, essentially of photovoltaic solar origin, are already competitive in a number of regions across the world. This trend shall continue and distributed generation should become competitive with retail electricity prices in most regions of the world in the coming decade (notably because the energy produced does not have to bear transmission and distribution costs). The potential of distributed generation, when looking at the actual surface of rooftops in residential homes and buildings which could be equipped with solar systems, exceeds the actual electricity consumption of the world by 2035. The potential of a significant increase in distributed generation is thus extremely important. Load defection is thus a potential reality which could disrupt the existing market and its current actors. According to all sources, in average, the volume of load defection should not however exceed 6% of the actual electricity distributed by 2030, with certain regions going as high as 10%. This seems to be a conservative forecast, although it is backed up by several sources. The coming years will tell if this competitive source of power will not unbalance electricity markets further. Additionally, distributed generation could lead to an accelerated substitution of fossil fuels used in buildings (and possibly in short-distance transportation) towards electric consumption, leading to both an increase in the share of electricity in the overall generation mix, as well as a decrease of the fossil fuel market for buildings' heating and transportation. The potential here is significant, and could lead all electricity consumption forecasts to be proven wrong by that time.

Finally, the evolution of the penetration of renewable energy may have an important impact on the way the grid operates. The ratio previously presented helps us better understand the dynamics of electricity markets and the possible evolution of retail electricity prices. However, it does not facilitate understanding of the additional stress renewable energies will create on the network and on the structure of the power generation market. The renewable penetration rate is expected to remain low in terms of energy consumed (TWh), due essentially to the low capacity factors at which those sources operate. Now, the production output of variable renewable energies (wind, photovoltaic solar) essentially varies in a day. At a certain time, they can indeed all operate together (a sunny and windy day for instance). The maximum rate of penetration of variable renewable energies must thus be evaluated to understand to which extent a grid can be put under stress at a given period of the day. Considering that all power generation from variable renewable sources could end up operating at their maximum power output at the same time, the maximum rate of penetration of variable renewable could reach new levels. According to the evaluations done, it could go on average from 15% today up to 31% (CPS scenario), 53% (450 scenario) and 139% (Energy [R]evolution scenario). Europe is today the most concerned grid with regards to variable renewable penetration as up to 41% of the total power generation could be supplied by these sources of energies at a certain time of the day already. Other regions range much below, with North America around 13%, China 14% and India 15%. This should strongly evolve in the coming years. Even in the very conservative CPS scenario, most regions should get closer to 30% by 2030 (North America, China, India, and OECD Asia), while Europe would reach around 71% of the total power generation consumed at a given time. The Energy [R]evolution scenario, which assumes a high penetration of variable renewable energies in the power generation mix, leads to almost all regions being way above 100% of their total power consumption. The progressive penetration of variable renewable energies is thus expected to lead to significant grid management issues. These issues will be examined in the next chapter.

As a final point, renewable energies are expected to take a very significant share in the power mix in the coming decades, with a part of it distributed to homes and buildings. The cost structure of renewable energies is that of base load units, with zero marginal cost of production. Consequently, renewable energies are likely to take the place of traditional base load units, with very low and very competitive prices. This represents a significant change of paradigm for power systems since their intermittent nature leads to a number of grid issues, which will be examined in the Chap. 4. Eventually, this evolution could yield an abundance of cheap electricity. Electrical energy (TWh) could even become free, consumers paying only for the capacity infrastructure to access it. This would deeply modify over time the way current markets operate, as well as create massive opportunities for productivity gains (and thus an increase in fuel switching to electricity in buildings, industry, and transportation). The distributed aspect of renewable energies could also yield major evolutions. Self-generation indeed presents the fantastic advantage to not require any more expensive transmission and distribution infrastructures. By nature, self-generation is thus an interesting economic alternative to grid energy. All forecasts tend to indicate that it will reach grid parity everywhere in the world in the coming years. The massive deployment of distributed generation (although not forecasted today) could lead to a complete redesign of the role of the grid in the coming decades. Instead of supplying the full amount of energy required by consumers, the grid infrastructure could evolve towards connecting "islands" of self-generation, or microgrids, in order to maintain balance and an overall high level of reliability across the grid. This could yield significant savings on grid infrastructure costs. However, this transition will not come at zero cost since important investments will have to be made to unlock its potential. In some countries, the decision will be obvious, because of the particularities of their current mix, competitive landscape, and consumption evolution. In others, it will require a clear political orientation. Paradoxically, depending on the mode of financing, short-term prices could end up increasing, except if this financing is borne by taxes. The fast deployment of renewable energies as base load generation will also create a number of grid issues, essentially related to their lack of flexibility, the intermittency of their operation, and their geographical distribution throughout the grid. These issues will be reviewed in the next chapter.

References

© OECD/IEA, Costs. (2015). *Projected costs of electricity.* https://www.iea.org/bookshop/711-Projected_Costs_of_Generating_Electricity
© OECD/IEA, Solar. (2014). https://www.iea.org/media/freepublications/technologyroadmaps/solar/TechnologyRoadmapSolarPhotovoltaicEnergy_2014edition.pdf
© OECD/IEA, Subsidies. (2016). http://www.worldenergyoutlook.org/resources/energysubsidies/
© OECD/IEA, WEIO. (2014). https://www.iea.org/publications/freepublications/publication/WEIO2014.pdf
© OECD/IEA, WEO. (2012). http://www.worldenergyoutlook.org/publications/weo-2012/
© OECD/IEA, WEO. (2014). *The energy [R]evolution scenario.* http://www.greenpeace.org/international/Global/international/publications/climate/2015/Energy-Revolution-2015-Full.pdf and https://www.iea.org/publications/freepublications/publication/WEO2014.pdf
AEMO. (2012). *Rooftop PV information paper.* Australian Energy Market Operator. http://www.aemo.com.au/Electricity/~/media/Files/Other/forecasting/Rooftop_PV_Information_Paper.ashx.
AER. (2016). Australian Energy Regulator. https://www.aer.gov.au/wholesale-markets/wholesale-statistics/weekly-volume-weighted-average-spot-prices
ANEEL. (2016). http://www.aneel.gov.br/area.cfm?idArea=550
BNEF. (2017). *Henbest: Energy to 2040 – faster shift to clean, dynamic, distributed.* https://about.bnef.com/blog/henbest-energy-2040-faster-shift-clean-dynamic-distributed/
BNEF/NEO. (2017). New Energy Outlook. https://about.bnef.com/new-energy-outlook/
BNEF/Solar. (2017). *Solar could beat coal to become the cheapest power on earth.* https://www.bloomberg.com/news/articles/2017-01-03/for-cheapest-power-on-earth-look-skyward-as-coal-falls-to-solar
BOI. (2016). http://www.boi.go.th/index.php?page=utility_costs&language=en
Brazil Business. (2016). http://thebrazilbusiness.com/article/electricity-prices-in-brazil
Business Insights. (2010). Breeze, P. *Costs of power generation.* http://lab.fs.uni-lj.si/kes/erasmus/The%20Cost%20of%20Power%20Generation.pdf
Business Tech. (2016). http://businesstech.co.za/news/energy/99494/electricity-prices-south-africa-vs-the-world/
California Photon. (2010). http://californiaphoton.com/energy/world/rooftop.html
Chadbourne. (2014). *Solar PV (DG).* http://www.chadbourne.com/rooftop_solar_june2014_projectfinance
Credit Suisse. (2015). *Solar outlook (DG).* https://doc.research-and-analytics.csfb.com/docView?language=ENG&format=PDF&source_id=csplusresearchcp&document_id=1043033521&serialid=Nf0NvzFluTA%2BQvDV1smDThFacN5IgqEVyQVK6QSJDHY%3D

Deloitte. (2015). *U.S. Solar power growth through 2040.* http://www2.deloitte.com/content/dam/
    Deloitte/us/Documents/energy-resources/us-er-solar-innovation-growth.pdf
Deloitte/Capacity. (2015). *The future of the global power sector.* Preparing for emerging opportu-
    nities and threats. https://www2.deloitte.com/content/dam/Deloitte/global/Documents/Energy-
    and-Resources/gx-power-future-global-power-sector-report.pdf
DEWA. (2016). http://www.dewa.gov.ae/tariff/newtariff.aspx
EIA Price. (2016). http://www.eia.gov/electricity/monthly/epm_table_grapher.cfm?t=epmt_5_03
EIA/Capacity. (2016). *International energy outlook.* http://www.eia.gov/forecasts/aeo/data/
    browser/#/?id=16-IEO2016&region=0-0&cases=Reference&start=2010&end=2040&
    f=A&sourcekey=0
EIA/Power Generation. (2015). https://www.eia.gov/forecasts/aeo/electricity_generation.cfm
EIA/Wholesale. (2016). *Wholesale prices.* http://www.eia.gov/electricity/wholesale/
European Commission/Wholesale. (2016). *Quarterly report on European electricity markets.*
    https://ec.europa.eu/energy/sites/ener/files/documents/quarterly_report_on_european_electric
    ity_markets_q4_2015-q1_2016.pdf
EWB. (2016). http://www.ewb.ch/de/angebot/strom/privatkunden/preise/home.html
Fraunhofer.   (2013).   https://www.ise.fraunhofer.de/en/publications/veroeffentlichungen-pdf-
    dateien-en/studien-und-konzeptpapiere/study-levelized-cost-of-electricity-renewable-energies.
    pdf
Global CCS Institute. (2015). http://www.globalccsinstitute.com/insights/authors/LawrenceIrlam/
    2015/07/24/levelised-costs-electricity-ccs
Global Climate Scope. (2016). http://global-climatescope.org/en/country/nigeria/#/details; http://
    global-climatescope.org/en/country/pakistan/#/details
Green Rhino Energy. (2016). http://www.greenrhinoenergy.com/renewable/context/economics.php
Greenpeace. (2015). *The energy [R]evolution scenario.* http://www.greenpeace.org/international/
    Global/international/publications/climate/2015/Energy-Revolution-2015-Full.pdf
Hong Kong Electric. (2016). https://www.hkelectric.com/en/customer-services/billing-payment-
    electricity-tariffs/domestic-tariff
HOTG. (2006). http://heartsofthegods.blogspot.ch/2006/11/giant-american-houses-another-symp
    tom.html
IISD. (2016). International Institute for Sustainable Development. https://www.iisd.org/gsi/renew
    able-electricity-subsidies
ILAR. (2012). ILAR Working Paper #7. *What does it cost to build a power plant?* http://ilar.ucsd.
    edu/assets/001/503883.pdf
Institute for Energy Research. (2015). *The levelized cost of electricity from existing generation
    resources.* http://instituteforenergyresearch.org/wp-content/uploads/2015/06/ier_lcoe_2015.pdf
IPCC. (2011). Renewable energy sources and climate change mitigation special report. *Intergov-
    ernmental panel on climate change.* http://srren.ipcc-wg3.de/report
Irena/Solar.   (2012).   International   renewable   energy   agency.   http://www.irena.org/
    DocumentDownloads/Publications/RE_Technologies_Cost_Analysis-SOLAR_PV.pdf
JEPX. (2016). Japan Electric Power eXchange. http://www.jepx.org/english/index.html
Kalam, A., King, A., Moret, E., Weerasinghe, U., & Roytburd, M. (2009). *Combined heat and
    power systems: Identifying economic and policy barriers to growth.* http://franke.uchicago.edu/
    bigproblems/Team5-1209.pdf
Kepco. (2016). http://cyber.kepco.co.kr/ckepco/front/jsp/CY/E/E/CYEEHP00201.jsp
Leonardo. (2007). *Life expectancy of nuclear power plants.* http://www.leonardo-energy.org/blog/
    life-expectancy-nuclear-power-plants
Lumo   Energy.   (2016).   https://lumoenergy.com.au/cms/images/pdf/20130701_pfs_sa_elec_
    businessoptions.pdf
Manitoba Hydro. (2016). https://www.hydro.mb.ca/regulatory_affairs/energy_rates/electricity/util
    ity_rate_comp.shtml

MB. (2016). http://www.mb.com.ph/for-sixth-straight-month-power-rates-down-in-october-meralco/#R9eXOH8skSMhUIGO.99

McKinsey. (2012). *Darkest before dawn. (DG)*. https://www.mckinsey.com/client_service/sustain ability/latest_thinking/~/media/5E847C563A734F148B5F3A6EFBD46E39.ashx.

NREL. (2010). http://www.nrel.gov/docs/fy11osti/48595.pdf

NREL/Coal. (1999). *Lifecycle assessment of coal fired power production*. http://www.nrel.gov/docs/fy99osti/25119.pdf

NREL/Projections. (2015). *Renewable electricity: Insights for the coming decade*. http://www.nrel.gov/docs/fy15osti/63604.pdf

NREL/Solar. (2015). *Standard scenarios annual report: Sector scenario exploration*. http://www.nrel.gov/docs/fy15osti/64072.pdf

Ovo Energy. (2016). https://www.ovoenergy.com/guides/energy-guides/average-electricity-prices-kwh.html

Parkinson, G. (2015). *Rooftop solar to overtake coal capacity in Australia by 2030*. http://reneweconomy.com.au/2015/rooftop-solar-to-overtake-coal-capacity-in-australia-by-2030-2030

PLN. (2016). http://www.pln.co.id/wp-content/uploads/2015/05/Tariff-Adjustment-Juni-2015.pdf

PSI. (2014). Paul Sherrer Institut. *Review of Swiss electricity scenarios 2050*. https://www.psi.ch/eem/PublicationsTabelle/PSI-Bericht_14-05.pdf

PVEducation. (2016). *Average solar radiation*. http://www.pveducation.org/pvcdrom/properties-of-sunlight/average-solar-radiation

REN21. (2013). http://www.ren21.net/Portals/0/documents/activities/gfr/REN21_GFR_2013.pdf

Roland Berger, (2015). Solar PV (DG). https://www.rolandberger.com/media/pdf/Roland_Berger_TAB_Solar_PV_20150610.pdf

Saudi Electric. (2016). https://www.se.com.sa/en-us/customers/Pages/TariffRates.aspx

Science Direct. (2014). *Techno-economic analysis of solar photovoltaic power plant for garment zone of Jaipur city*. http://www.sciencedirect.com/science/article/pii/S2214157X13000117

Shenzhen. (2016). http://english.sz.gov.cn/ln/201205/t20120517_1914423.htm

Singapore Power. (2016). http://www.singaporepower.com.sg/irj/servlet/prt/portal/prtroot/pcd!3aportal_content!2fcom.sp.SPEPCustomization!2fcom.sp.SP_Default_Framework!2fcom.sp.SP_Custom_Desktop!2fframeworkPages!2fcom.sp.SP_Custom_Default_Framework_Page!2fcom.sap.portal.innerpage?windowId=WID1332135032094&NavigationTarget=navurl://41c8e6a3faf48bb168af2c222faa8ee4&windowId=WID1332134938209

Standard & Poor's. (2014). *Dow Jones Industrial Average*. Dow Jones Indices-McGraw Hill Financial. www.djindexes.com/mdsidx/downloads/fact_info/Dow_Jones_Industrial_Average_Fact_Sheet.pdf.

Statista/Capacity. (2016). *Projected additions to power generation capacity from 2010 to 2030*. http://www.statista.com/statistics/498310/added-power-generation-capacity-worldwide-fore cast-by-energy-type/

Storm Report. (2016). http://strom-report.de/strompreise/#strompreise-europa

Sudan Tribune. (2016). http://www.sudantribune.com/spip.php?article45028

Synapse Energy. (2016, Spring). *National carbon dioxide price forecast*. http://www.synapse-energy.com/sites/default/files/2016-Synapse-CO2-Price-Forecast_0.pdf

Telegraph/Wind. (2016). *Wind farm turbines wear sooner than expected*, says study. http://www.telegraph.co.uk/news/earth/energy/windpower/9770837/Wind-farm-turbines-wear-sooner-than-expected-says-study.html

Thomson Reuters. (2014). *Point carbon. The MSR: Impact on market balance and prices*. http://ec.europa.eu/clima/events/docs/0094/thomson_reuters_point_carbon_en.pdf

Times of Israel. (2016). http://www.timesofisrael.com/electricity-prices-to-drop-by-9-4-in-2015/

TNB. (2016). https://www.tnb.com.my/residential/pricing-tariffs

TurkStat. (2016). http://www.turkstat.gov.tr/PreHaberBultenleri.do?id=15883

UK DECC. (2015). Department of Energy and Climate Change. https://www.gov.uk/government/uploads/system/uploads/attachment_data/file/419024/DECC_LowCarbonEnergyReport.pdf

Vietnam News. (2016). http://vietnamnews.vn/economy/265858/power-tariffs-to-rise-95-per-cent.html

Vuorinen, A. (2009). *Optimal power systems*. www.ekoenergo.fi/page27.php

World Bank/Carbon Pricing. (2015). http://www-wds.worldbank.org/external/default/WDSContentServer/WDSP/IB/2015/08/26/090224b08309a09a/4_0/Rendered/PDF/Carbon0pricing0e0released0late02015.pdf

World Energy Council. (2013). *World energy perspective. Cost of energy technologies*. https://www.worldenergy.org/wp-content/uploads/2013/09/WEC_J1143_CostofTECHNOLOGIES_021013_WEB_Final.pdf

Chapter 4
The Transition to a New Grid

4.1   A Model for Understanding Grid Challenges: The "Duck" Curve

The massive upcoming penetration of renewable into the overall power generation mix, in particular variable renewable energies such as wind and photovoltaic solar, creates new issues which will force a redesign of the way a grid operates, both from a market and a technical standpoint.

4.1.1   Main Issues Associated with High Share of Variable Renewable Energies

Variable renewable energies are by nature intermittent. Their production outputs vary from one day to another, one season to another. The balance of power (also called net load) must thus be adjusted to their actual production output in real time (World Economic Forum 2015). This leads to periods of the day where the production output of renewable energies is very high, and periods where it is extremely low. During the time renewable energies operate close to their maximum output, other power sources must see their production output reduced. Some of these sources enjoy a high capacity factor, with some operating continuously on the network. The progressive emergence of variable renewable energies, which cannot be regulated, yields a reduction of the capacity factor of these units, and some need to be shut down part of the day. This drastically modifies the economics of these units, which consequently receive less revenue to compensate for their (fixed) costs. Flexible units such as natural gas power plants are better prepared for this as the share of their variable costs is indeed more important. Now, less flexible units, for instance nuclear power plants, cannot accept a drop in their capacity factor below a certain limit. "Minimum load balancing" corresponds to the amount of supply to be provided by

© Springer Nature Switzerland AG 2019
V. Petit, *The New World of Utilities*, https://doi.org/10.1007/978-3-030-00187-2_4

conventional generation when renewable energies are considered to be operating at their maximum output (© OECD/IEA, Power Transition 2012). The volume of supply will then vary over the day, depending on the actual output of renewable energies. A necessary amount of conventional technologies must thus be kept available on the network for those periods of time when renewable energies do not operate, or operate below their expected level ("maximum load balancing"). We will return to the financial implications of such a requirement later on.

Although the typical production output of a wind farm does not depend much on the time of the day, but rather on weather conditions, photovoltaic solar farms are dependent on the irradiation level, thus the time of the day. Solar based energies tend to operate more efficiently at midday, with performance dipping at the end of the afternoon. Demand is not homogeneous during the day, with power consumption moving between a low and a high. The load curve helps to visualize the actual evolution of power consumption throughout the day. There is traditionally a "peak" of consumption at the end of the day, which corresponds roughly to the time when people go home after work. This peak can be defined in absolute value, as well as relatively to the average load consumed by the users of the grid. The latter ratio is called the "peakiness" of the grid. Grids with high peakiness tend to experience more stress on the network and on generation units since the variations of the load are greater. This peak typically occurs at a time when renewable production output is low, which means that conventional technologies must pick up and supply the necessary amount of power to meet the peak demand level (© OECD/IEA, Capacity Credit 2011). The more variable renewable energy on the network, the lower the production output of conventional technologies at midday. This leads to an important and accelerated ramp up of conventional technologies towards the end of the day in order to supply the peak. This also corresponds to an artificially increased "peakiness" of the grid. This ramp up is not easy to realize as most conventional technologies do not have the sufficient flexibility to be shut down and restarted several times during the day. Such shutdowns and restarts increase wear and tear of these units in the long run.

The output of variable renewable energies is also difficult to predict. In the United Kingdom, forecasting errors have indeed jumped from 7% before the introduction of renewable energies to 28% today (© OECD/IEA, Networks Infrastructure 2013). This uncertainty of their actual production partially disrupts the traditional market organization. Indeed, additional "reserves" must be considered in order to cope with the uncertainty of renewable supply. Consequently, an increased number of balancing operations must be decided, leading to an increased cost associated with balancing the network with the unpredictability of renewable supply (© OECD/IEA, Power Transition 2012).

Renewable energies have no inertia capability since they are not synchronized machines over the grid. Photovoltaic solar farms produce DC current which is converted through inverters into AC current that can be supplied over the grid. Consequently, those power sources do not constitute rotating machines synchronized over the grid. In case of an additional (or brutal drop of) load demand, synchronized machines tend to "resist" the effort. This resistance is called inertia.

It is inertia which helps maintain the frequency level and the stability of the network. Renewable sources do not have this capability and thus cannot support the stability of the network. The more the variable renewable energies over the network, the less the volume of synchronized machines with inertia for a given power demand. Ancillary services must thus be reinforced on a grid with a high share of renewable in order to control the frequency in real time in an effective manner (Enbala 2013).

Finally, the penetration of renewable energies is not geographically homogeneous. Typically, wind farms are installed in high-wind areas, close to sea coasts for example. Solar farms are installed in areas with high irradiation. The selectivity of the areas where renewal power capacities can be installed leads to local imbalance and network congestion, as well as increased costs associated with the transportation of electricity to areas of consumption.

4.1.2   The "Duck" Curve

In summary, the penetration of a high share of renewable energies imposes a number of network issues. First, it modifies the actual mix of power generation units on the grid by reducing the minimum load balancing level, at best from a capacity factor standpoint and, over time, by removing conventional technologies which cannot be profitable anymore. Then, it leads to increased stress over the network by artificially modifying the "peakiness" of the network, with accelerated ramp up and down of conventional units. It also leads to increased costs because of the necessary balancing services (forecast errors) and ancillary services (frequency control) to control the grid, as well as increased transmission and distribution costs to manage local congestion and network issues, and the losses related to the transportation of electricity to areas of mass consumption.

A model can be formalized which depicts the various issues a grid can experience with the introduction of variable renewable energies (Fig. 4.1). The maximum production output of renewable energies leads at certain times of the day to price issues (low price levels at certain periods of the day due to the zero marginal cost of renewable energies) and capacity factor issues (traditional power sources not operating or operating at low load). This reduces the minimum load balancing set point, and over time reduces the conventional power plants' footprint to the profit of renewable energies. The minimum penetration rate of renewable leads however to scarcity of power at certain times of the day as well as inertia issues, which creates peak capacity and load balancing constraints. Finally, the localization of variable renewable energies leads to increased transmission and distribution costs.

Another way of looking at the issues associated with the penetration of variable renewable energies is to look at the net load curve evolution with the share of renewable energies. The California Independent System Operator (2013) load curve evolution is presented in Fig. 4.2. The 2012 load curve shows a variation between 18 GW and 23 GW during the day. The peak load is evolving upwards to 26 GW by 2020. The penetration of variable renewable energies leads to a strong

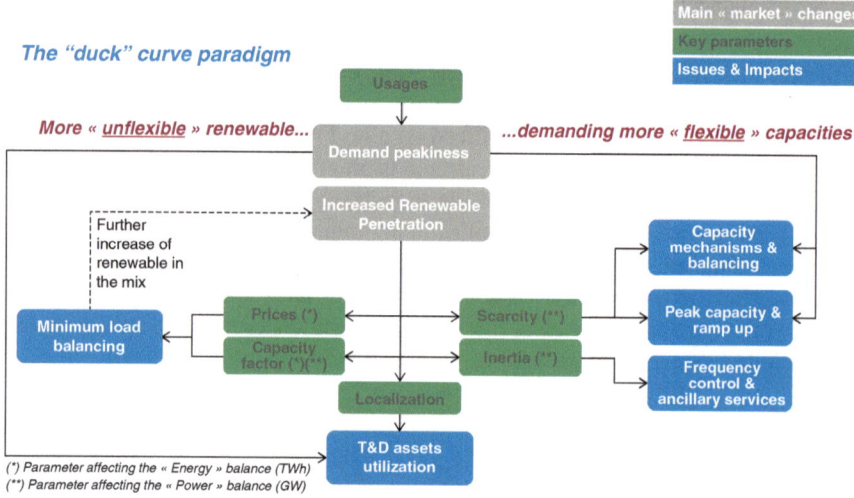

Fig. 4.1 The new grid challenges

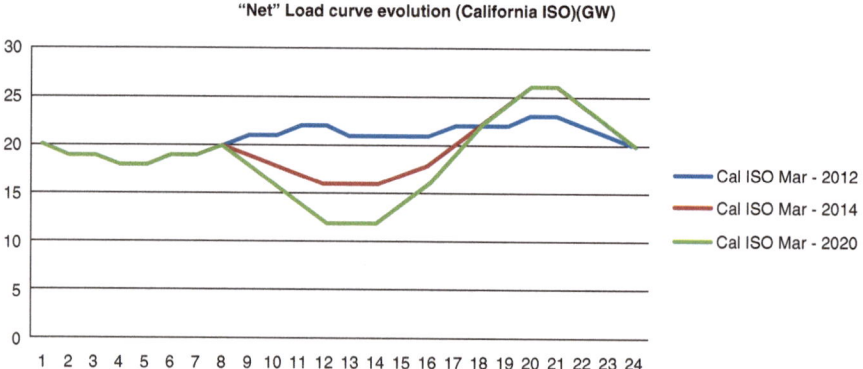

Fig. 4.2 Net load curve (duck curve) evolution (California ISO 2016)

drop of the net demand for conventional power generation at midday. The net load curve depicts this evolution by subtracting the renewable output from the actual load curve. The minimum load balancing drops at 12 GW, leading to 6 GW of power not operating continuously on the network anymore. This represents more than 25% of the actual power capacity demand at any time of the day. Additionally, more than 10 GW of power needs to be started up in 5 h to meet "peak" demand, when less than 1 GW was required before, leading to a complete change of paradigm on how to operate the various generation units on the network. Finally, the uncertainty of the production outputs of renewable energies on the network can lead to a variety of situations. There is indeed more than 10 GW of wind and solar energies connected to the network today for a total capacity of 78 GW (Energy Almanac 2014). Around half of this power capacity is considered in the load curve presented in Fig. 4.2. Unpredictable weather conditions (good or bad) could lead this amount of capacity to be higher or lower, leading conventional generation output to adjust accordingly. The uncertainty of variable renewable production output thus leads to an increase in balancing services from conventional units. Finally, the lack of inertia from renewable is not visible here. With a minimum load balancing of 12 GW compared to 22 GW, the inertia capability of the overall set of power generation is significantly reduced in the example of California, leading to a more unstable network, and consequently to additional ancillary services having to be planned and paid for during the time renewable energies operate. The 2020 California ISO load curve is also called the "duck" curve because of its specific form which resembles a duck and is a good description of most issues which apply to the grid with a high share of variable renewable energies.

To sum up, the overall emergence of the "duck" curve paradigm is a major change to how grids and their corresponding markets are being operated. The massive influx of non- flexible renewable energies on the network creates cascading needs for a greater flexibility from other actors in the network.

4.1.3   A Model to Enable Quantitative Analysis

The "duck curve" is a proper illustration of the issues affecting a grid with an increasing share of renewable. The corresponding "duck curves" are not easily retrievable, especially when it comes to forecasting their evolution over time. However, they can be estimated based on a number of assumptions.

First, the load curve can be retrieved for many countries and regions. The following load curves were retrieved:

– California (California ISO 2016; California ISO/Load Curve 2016)
– Texas (ERCOT 2016)
– France (RTE 2016)
– Spain (Red Electrica 2016)
– Germany: the load curve is retrieved by consolidating the load curves of the four main transmission operators (Amprion 2016; Tennet 2016; 50 Hertz 2016; Transnet BW 2016).
– Nordics (Nordpool 2016; Statnett 2016)
– Turkey (TEIAS 2008; WWF 2014)
– New South Wales, Australia (Transgrid 2014).

The actual production capacities (both renewable and total) can also be retrieved from various sources (California ISO/Renewable 2016; ERCOT/Renewable 2016; RTE/Renewable 2016; Red Electrica/Renewable 2013; Clean Energy Wire 2016; Fraunhofer/Renewable 2016; Statnett 2016; WWF 2014; Transgrid/Renewable 2016).

Fraunhofer (2016) can be used as a model to map the effect of renewable energies on the load curve. The study of the various load curves of Germany shows photovoltaic solar operates on average at 59% of its rated capacity in summer (S) and 33% in winter (W). Depending on the day, this ratio can vary between 53% and 66% in summer, and 13–53% in winter. Photovoltaic solar is also not available continuously throughout the day. It follows a bell curve with 9–10 h of load around midday. Wind power operates differently. It does not depend on the time of the day but only on the wind situation over the region where it operates. According to Fraunhofer (2016), the average observed wind power was around 26% of the rated capacity in summer (with a variation between 17% and 36%), and 56% in winter (with a variation between 47% and 64%).

A model of renewable penetration can be drawn from the analysis above (Fig. 4.3). This model is obviously highly theoretical, but it is possible with it to compute the "net load" or "duck" curve of the corresponding regions analyzed, by subtracting the renewable contribution to the actual load curve.

The forecasted evolution of the "duck" curve is also interesting as it enables the projection of the impact of an increasing share of renewable energy on the grid. The power consumption evolution can be retrieved by region (© OECD/IEA, WEO 2014; Greenpeace 2015) and is applied to each country or region of reference to simulate the actual increase in consumption in the region when no other data is available (data was retrieved for France, Germany, Texas and New South Wales). The corresponding load curve is retrieved considering the usages are unchanged, which means the profile of the load curve remains also unchanged. Here again, this simulation is theoretical since the variation of consumption can evolve within a region itself. The form of the load curve is also likely to evolve, depending on the evolution of usages. This assumption constitutes a best estimate, given that the projections of usages and demand evolution are extremely difficult to retrieve for up to 2030 at this level of granularity.

From this forecasted load curve, the projected "duck" curve can also be computed. To do this, the forecasted capacity of variable renewable energies (wind and photovoltaic solar) to be installed by 2030 needs to be first evaluated. These forecasts can be retrieved from various sources. The scenarios from the International

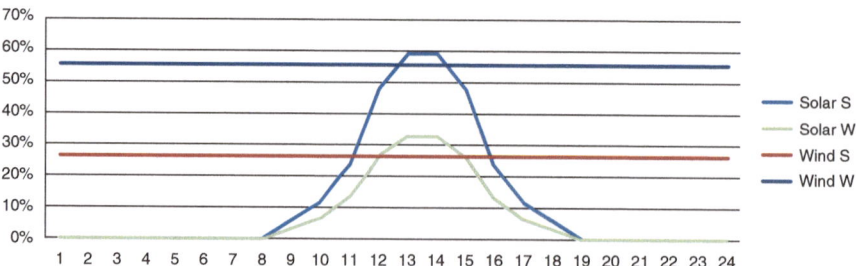

Fig. 4.3 Average contribution of solar and wind power to total capacity (Source: Author's basic model used in reference)

Energy Agency (© OECD/IEA, WEO 2012, 2014) and from Greenpeace (Greenpeace 2015) provide such forecasts for each region. The regional forecasts can be used to retrieve the evolution at the country/state level as a proportion of the regional one. This is a highly theoretical projection as obviously the rate of growth will vary within a region itself and between countries and states. The actual history of renewable energies' development can also be used to project the long-term evolution, assuming a continuous rate of growth over the next 20 years. Finally, this data can be compared to "other" forecasts which were gathered as elements of comparison. A final "selected forecast" is then issued from these comparisons (Figs. 4.4 and 4.5). This is the one which will be used further on.

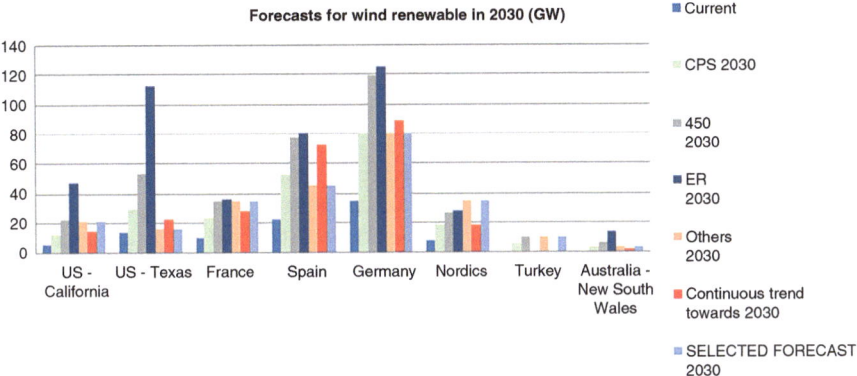

Fig. 4.4 Forecasts for wind renewable in 2030 (ADEME 2016; ERCOT/Forecast 2016; Irena 2014; McKinsey/Forecast 2016; NSW government 2015; pv magazine 2015a, b; STORE project 2016; Wind Power Monthly 2015; WWF 2014)

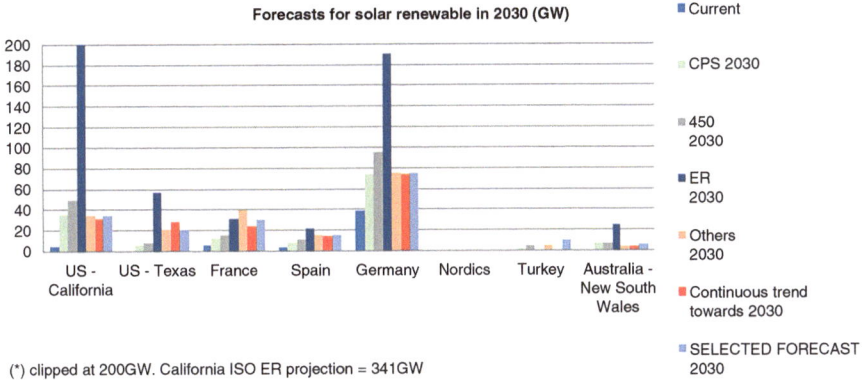

(*) clipped at 200GW. California ISO ER projection = 341GW

Fig. 4.5 Forecasts for solar renewable in 2030 (ADEME 2016; ERCOT/Forecast 2016; Irena 2014; McKinsey/Forecast 2016; NSW government 2015; pv magazine 2015a, b; STORE project 2016; Wind Power Monthly 2015; WWF 2014)

- U.S.—California: the RPS (Renewable Portfolio Standard) from California aims at reaching 50% of power issued by renewable, which would lead, depending on the assumptions, to 21 GW of wind and 34 GW of solar by 2030 (Wind Power Monthly 2015). This more or less corresponds to the continuation of the current rate of growth until 2030, as well as the 450 projection. This is the forecast that was selected.
- U.S.—Texas: the long-term assessment of ERCOT yields a small increase in wind (from 14 GW today to around 16 GW) but a very large increase in solar (from 1 GW to 21 GW) by 2030 (ERCOT/Forecast 2016). Globally, this is much lower for wind than any forecasts which were retrieved or evaluated, and more or less corresponds to a continuous trend for solar. This is the forecast which was selected.
- France: the forecast for wind retrieved from the European Wind Energy Association (2015) yields 35 GW of wind power by 2030 (compared to 10 GW today). This is lower than estimates from the ADEME (2016) of around 60 GW in their high renewable scenario, but very close to the projections from both the 450 and the Energy [R]evolution scenarios, as well as the continuous trend. 35 GW was thus finally considered for wind. The ADEME projections yielded around 40 GW for photovoltaic solar by 2030. These projections are higher than all other projections. It was decided to select a lower forecast of 30 GW (closer to the continuous trend).
- Spain: the STORE project (2016) has estimated wind and solar forecasts to reach 60 GW for wind and 16 GW for solar (middle scenario, half completion of the 80% scenario by 2030). The European Wind Energy Association (2015) has however estimated the wind forecast to not reach above 45 GW by 2030 (compared to the target of the country of 35 GW by 2020). The selected forecast included 45 GW for wind (a bit lower than other projections) and 16 GW for solar (similar to the continuous trend of development).
- Germany: the European Wind Energy Association (2015) has estimated wind power to reach 80 GW by 2030 (compared to 35 GW today). This corresponds to the forecast from the STORE project (2016) at around 86 GW as well as projections from the continuous trend and the CPS scenario. 80 GW is selected here for wind power. As for solar, various forecasts estimate it to reach 75 GW by 2030 (pv magazine 2015a; STORE project 2016). This corresponds as well to the continuous trend as well as the CPS scenario. This is the forecast selected here.
- Nordics: the European Wind Energy Association (2015) has proposed a forecast of around 35 GW for Nordics overall (Denmark, Sweden, Finland) compared to 8 GW today. This is more aggressive than most projections and was retained here as the final forecast. As for solar, the development is more limited, and available forecasts estimated no more than 1 GW by 2030 (Irena 2014). This is what was selected here as well.
- Turkey: the WWF (2014) report estimates that wind power could reach 10 GW in Turkey by 2030. This fits well projections from the 450 scenario and was selected here. Various forecasts considered 5 GW by 2023 could be reached for solar

(pv magazine 2015b). This forecast corresponds to the projections from the 450 scenario as well, but is quite conservative considering the potential of the country. Consequently, the 2030 selected forecast used here was lifted up to 10 GW.

- Australia—New South Wales: assuming a similar ratio of renewable penetration for the state than for the whole country, various forecasts (McKinsey/Forecast 2016; NSW Government 2015) yield around 4 GW for wind and 4 GW for solar by 2030, out of 21 GW total capacity estimated by 2030. The Renewable Action Plan from the New South Wales government (2015) also yields 4 GW of distributed solar by 2030. The total 4 GW of solar by 2030 was thus lifted up to 6 GW to take into account a possible faster penetration rate of renewable energies, which recent years have so far confirmed.

These estimates enable a more quantitative analysis of the impact of renewable penetration on various grids around the world. Following this model, the increase in the share of renewable energies on the grid leads to a number of issues, among them minimum load balancing, "peak" load management, capacity balancing, real-time frequency control, and additional grid costs. The impact of those issues can then be analyzed quantitatively. While a highly theoretical exercise, it still represents the first quantitative analysis and forecast across countries.

4.2 New Grid Challenges

4.2.1 Structural Evolution of Conventional Generation

The massive penetration of renewable energy into the grid leads to an evolution of the minimum load balancing, the load supplied by conventional generation. If the traditional load curve evolves slightly around its average, the minimum load balancing corresponds to the minimum load supplied continuously during 1 day. This corresponds to base load units, which end up operating continuously. Nuclear power plants are typically used as base load, since their cost of electricity essentially depends on the capacity factor and on the cost of investment (an incentive to produce as much as possible to be profitable) and since the cost and the complexity associated with turning off and on a nuclear power plant is extremely high. Other conventional power plants such as natural gas (simple cycle) and coal are more easy to operate and their cost of production is more dependent on variable costs (operation and maintenance, fuel costs), which then vary as a function of their production output. Renewable energies such as wind and photovoltaic solar are also extremely dependent on the capacity factor. Their cost profile is that of base load units. They thus tend to operate as much as possible, and government regulations have traditionally enabled this. Now, their production output is not constant over 1 day and from 1 day to another, making it difficult to predict advance with accuracy. The remaining "net" load to be supplied by conventional generation thus needs to be adjusted accordingly. While this is not an issue as long as renewable energy is a minor contribution to the power supplied, this gets tricky as the volume of renewable capacities increases over the network. When renewable production gets very high, the capacity factor of conventional generation is decreased. This can lead to a decrease of the competitiveness of power sources, for which the cost of electricity is a function of the capacity factor. In the case of base load generation, this can also lead to more structural issues, with some plants not being able to operate anymore as the time and cost needed for them to ramp up/down their capacity on/from the grid is too long and too high to enable them to operate in such a variable environment. This leads them to not be usable anymore on the network, and this leads to a structural decrease of their capacity over the grid. This also creates an issue over the grid when renewable production is not operating anymore, as the load needs then to be picked up by other generation sources. With a structural decrease of the base load generation, this capacity must then be compensated by more flexible units like coal or gas (simple cycle) generation.

The minimum load balancing and the 2030 projection are mapped on Fig. 4.6 for various regions. First, the actual minimum load balancing can be compared to the actual average load curve. It shows that some countries such as Spain and Germany have a much lower minimum load than their average load consumption. For both countries, it has to do with the already important renewable penetration into the grid. Other regions have a minimum load balancing closer to the average load, hence a lower "peakiness". Then, the 2030 evolution varies significantly across countries. Indeed, the minimum load is both a result of the increased renewable penetration as well as of the increased consumption forecasted. In California, Spain and Germany, the minimum load balancing is expected to decrease by a significant factor, whatever

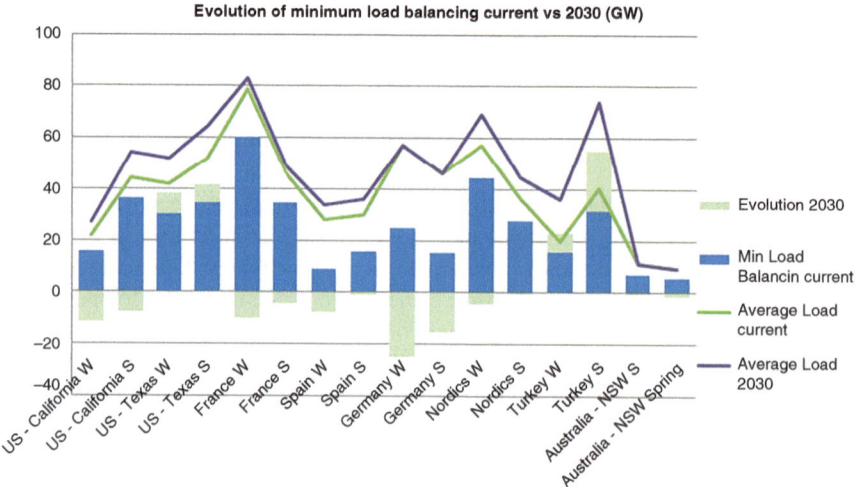

Fig. 4.6 Evolution of minimum load balancing (W: winter; S: summer) (Source: Author's own calculation)

the scenario, and even be zero in Germany. This would mean that any conventional power production would not be required anymore at certain periods of the day. In France, the Nordics and Australia, the minimum load balancing is reduced by a small factor while the actual average load increases. In other countries, the effect is nullified by the general increase in the average load (Texas, Turkey). The regions with the highest decrease of their minimum load are the ones which will require the highest level of restructuring of their market. Other countries would face no specific issue with their base load fleet, at least considering current 2030 forecasts of wind and solar power deployment.

When renewable electricity does not operate, at times of low production, the load to be balanced out by conventional generation is at its maximum. This typically happens at night, when solar production output is null. The maximum load balancing can thus be estimated.

Similar to the minimum load balancing, some countries stand out with a strong evolution of their maximum load balancing. The maximum load balancing is generally expected to increase, following the evolution of power consumption (Fig. 4.7). Turkey notably stands out with almost a doubling of its maximum load balancing. Several regions reviewed here benefit from a reduction of their maximum load balancing (thanks notably to the important share of wind, which continues to blow when solar is not producing anymore). The effect is specifically strong in Germany and France (in winter). There, the low increase in power consumption does not compensate the permanent impact of renewable penetration on the network.

Both maximum and minimum load balancing can then be subtracted to evaluate the overall amount of conventional generation that is required to operate discontinuously during the day on the network (mid-peak to peak load). It is an indication of how much conventional power generation must be kept on the grid, despite the

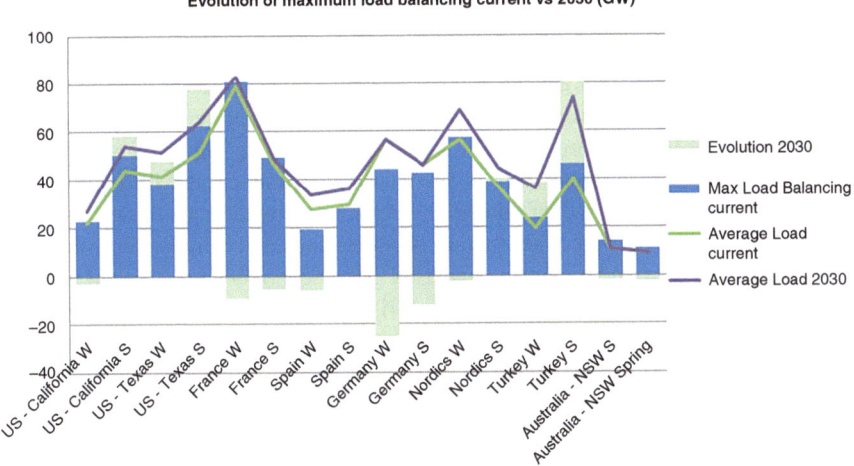

Fig. 4.7 Evolution of maximum load balancing (Source: Author's own calculation)

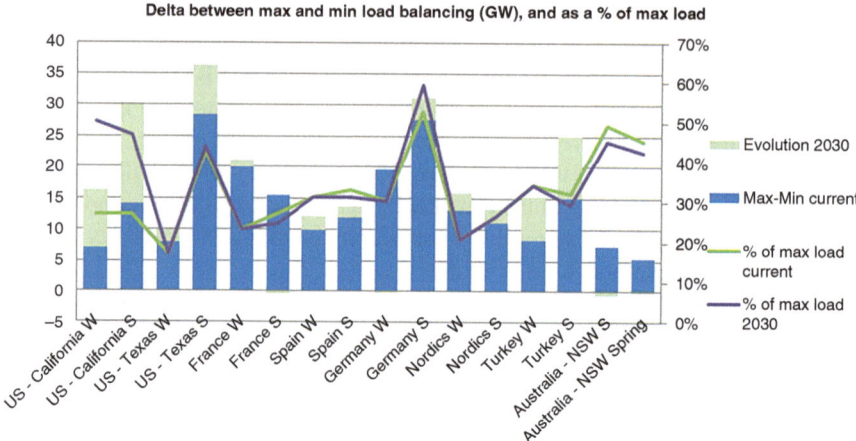

Fig. 4.8 Delta between maximum and minimum load balancing (Source: Author's own calculation)

penetration of renewable energies, in order to maintain electricity supply at all times, and in the absence of large storage technology available. It is presented in absolute value as well as a percentage of the maximum load (Fig. 4.8). California stands out as being extremely impacted by the penetration of renewable energies. There, the amount of mid-peak to peak load generation almost doubles. From an average of 30% of the maximum load, the volume of required intermittent conventional generation jumps to almost 50% of maximum load. This is mainly due to the significant evolution of solar output during the day, while wind is more constant (in our model). This difference is a major disruption on the power market as traditional generation needs to remain available to cope with the decrease of renewable output on the grid. In other regions, the power generation mix (in particular the renewable mix) makes this evolution less disruptive.

In summary, minimum load balancing is expected to decrease (or remain stagnant despite the increase in power consumption) in almost all mature economies. This is particularly true for California, Spain and Germany. There, the capacity factor of traditional conventional generation is decreasing, and this trend will accentuate in the coming years, leading to an unbearable economic equation for these generation sources. The base load generation market is thus expected to heavily restructure as it is substituted by variable renewable energies. New economies have less of an issue with this since their demand for electricity is rising significantly, leading to an offset of the impact of the introduction of renewable energies. There, renewable electricity is more thought of as a complement of power to meet rising demand. The maximum load balancing (maximum generation required from non-variable renewable sources at a given time in the day) is generally increasing everywhere, although tamed in mature economies, thanks to the high penetration of wind. The difference between maximum and minimum load balancing represents the actual volume of conventional power generation which shall remain available to cope with the maximum load requirement at any time of the day, while not being able to operate continuously (in absence of alternative storage technologies). This mid-peak to peak flexible power generation capacity generally ranges around 30% of the maximum load required at any time of the day. A proper economic equation needs to be confirmed for this type of operations. Mid-peak to peak flexible power generation capacity is expected to jump to 50% in California, a significant increase compared to the current situation. The power generation sources the least impacted by capacity factor variations are the most suitable for this type of operating model. Natural gas is typically the most appropriate, followed by coal power generation. In the last chapter, we will review the significant potential of emerging storage technologies for this application.

In conclusion, if the minimum load balancing is decreasing because of renewable energies, renewable power generation does not only substitute conventional power generation. Conventional power generation remains necessary to cope with those times when renewable does not operate. This volume of flexible supply (typically 30% of the maximum load) is expected to increase in certain regions. It should be an important problem for those regions, which would have to maintain high shares of conventional power generation to meet the variations in renewable output, at least in the absence of alternative (storage) technology.

According to Hirth (2012), the costs associated with the necessary flexibility in capacity are expected to increase strongly with higher penetration rates of renewable energies. They are essentially related to the decrease in capacity factor of conventional generation, which nevertheless is required to remain available to cope with periods of low renewable production output, and as such cannot be decommissioned. These costs are expected to be in the range of 15–35 Euro/MWh. This corresponds to a very significant share of the wholesale price of electricity (currently ranging around 50 USD/MWh in most countries), and these significant additional costs have to be taken into account.

4.2.2 The Problem of the "Peak"

The load curve, which reflects the demand in power, is not a constant over the day. It actually varies throughout the day. The daily demand then varies across usages. Consumers' usages determine the consolidated load demand that is required from the grid. The power consumption in a given country is a mix of residential, commercial, industrial and transportation consumption. The usages of each segment vary during the day. For instance, commercial buildings' power consumption is high during the time shops are open. Residential consumption is low during the day and at night, and tends to peak in the evening. Industrial consumption tends to be flat although it depends on the type of industry. Electric transportation varies depending on the mode of transportation. Railways and metro lines consume power when they operate, with a peak at times of high traffic, but electric cars consume only when plugged into the network, typically at night when the cars are not used. Figure 4.9 shows the evolution of the daily load profile of power consumption, depending on the time for different segments. The example comes from a study conducted in Brazil (Jardini et al. 2000).

Inside each of the segments, the usages may vary, leading to a variety of load profiles. Figure 4.10 shows the power consumption of residential usages for France (RTE/Usages 2014), Germany (Energy Research and Social Science 2014) and the United Kingdom (Intertek 2012). The study from Intertek is one of the most comprehensive studies done on electricity usages across Europe. It notably takes into account a variety of situations, including households with and without primary electric heating. The two power consumption usages are mapped in Fig. 4.10. The importance of space heating when performed by electric power is visible. It represents in France around 28% of the total electricity consumption and is extremely sensitive to weather conditions. In France, a 12° difference in temperature over a 2-week period in 2012 yielded a 30 GW difference in power consumption overall across the day, or more than 30% of the peak load observed for the country on the coldest day (above 100 GW)(RTE/Usages 2014).

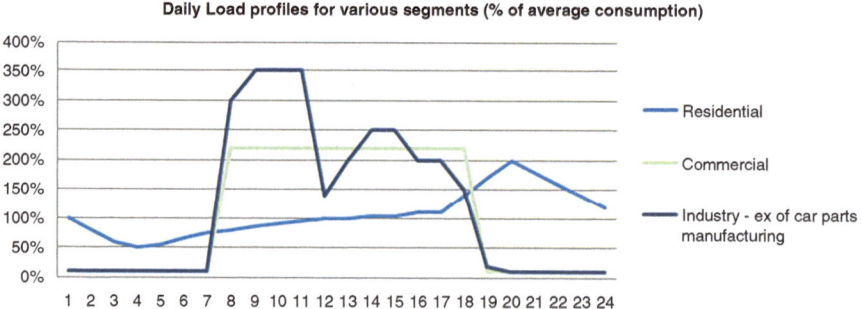

Fig. 4.9 Daily load profile for various segments (Jardini et al. 2000)

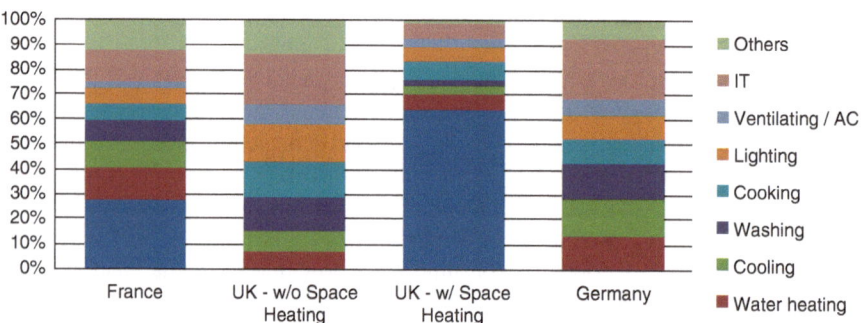

Fig. 4.10 Residential segment power consumption usages (Energy Research and Social Science 2014; Intertek 2012; RTE/Usages 2014)

Now, each type of electricity usage does not occur throughout the day. Typically, space heating and cooking are required when the inhabitants are at home. Water heating can be regulated across the day. Lighting is required at night. IT will depend on usages at home. The load profile throughout the day will thus depend on people usages. Intertek (2012) conducted a study in the United Kingdom of the detailed load profiles for a variety of households. It segregated households by type, for example, houses with and without electric heating (Figs. 4.11 and 4.12). The load profile of households without electric heating shows a peak of power consumption around the end of the day because of lighting and cooking that happen at that part of the day. In households with electric heating, space heating is more important at night and the peak actually occurs overnight. Power consumption in such households is also considerably higher than in those without electric space heating.

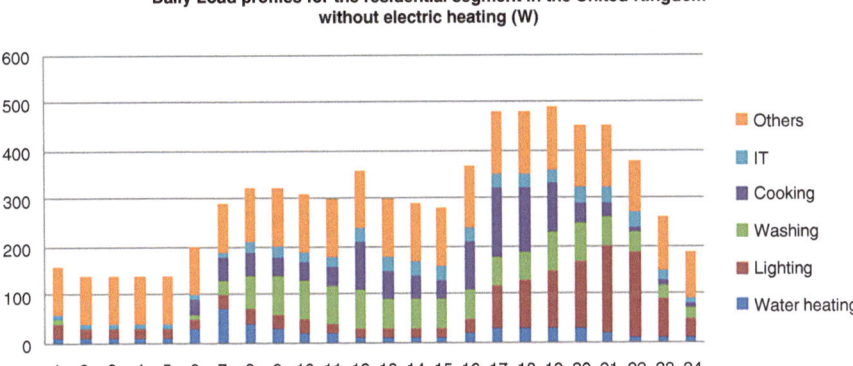

Fig. 4.11 Daily load profiles for the residential segment in the UK—no electric heating (Intertek 2012)

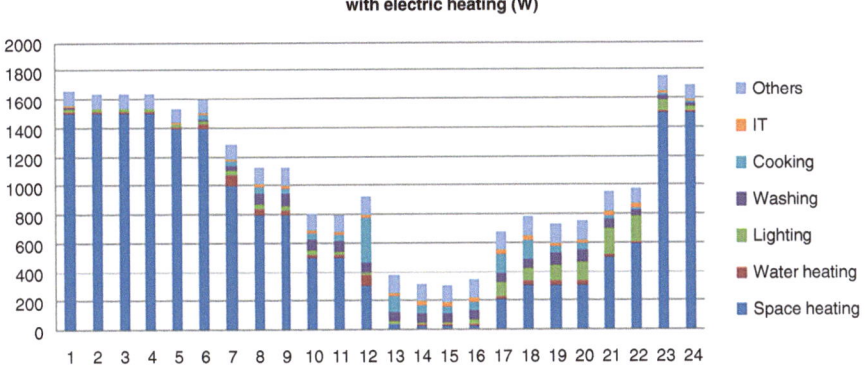

Fig. 4.12 Daily load profiles for the residential segment in the UK—electric heating (Intertek 2012)

In Australia (Energymag 2016), the daily profile varies between winter and summer (Figs. 4.13 and 4.14). Water heating represents a very high share of the total power consumed. It operates continuously during the day, both in winter and summer, consumes more electricity at night, with higher usage in winter than in summer. During winter, the peak at the end of the day comes from the cumulated consumption of heating, lighting and IT. In summer, the load profile is different as air conditioning progressively kicks in during the day, leading to a progressive increase in power consumption up until peak time around 6 pm.

The different individual load profiles add up to the overall demand to be supplied. These load profiles are different from one region to another, as shown in Fig. 4.15. Nevertheless, the mapping of 11 utilities at different times of the year shows a certain continuity in the shape of the load profile. The profile is low in the morning hours. A first ramp up occurs around 6–7 am, when people wake up and begin commuting.

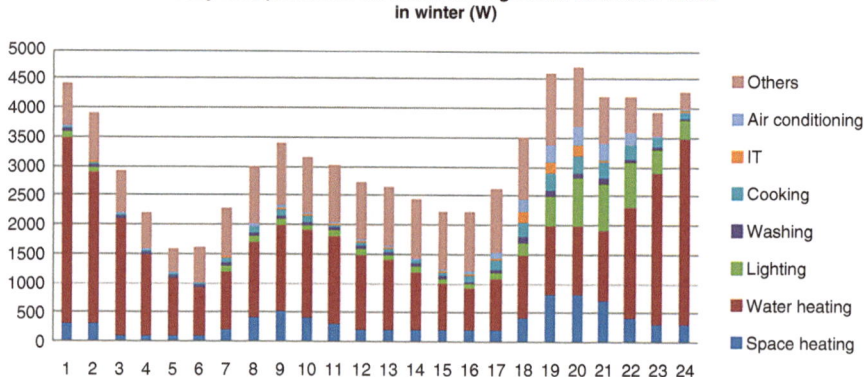

Fig. 4.13 Daily load profiles for the residential segment in New South Wales—winter (Energymag 2016)

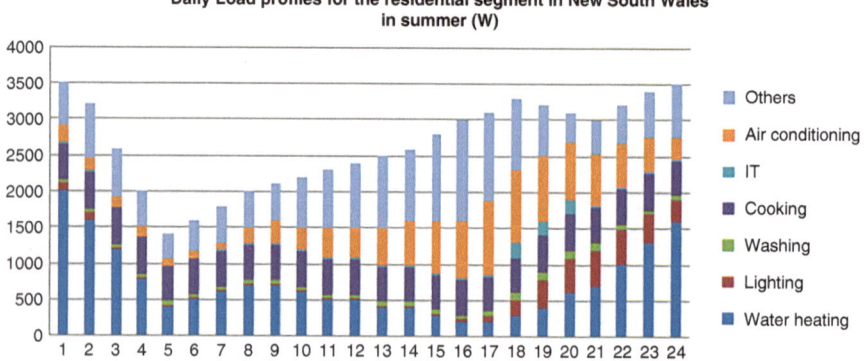

Fig. 4.14 Daily load profiles for the residential segment in New South Wales—summer (Energymag 2016)

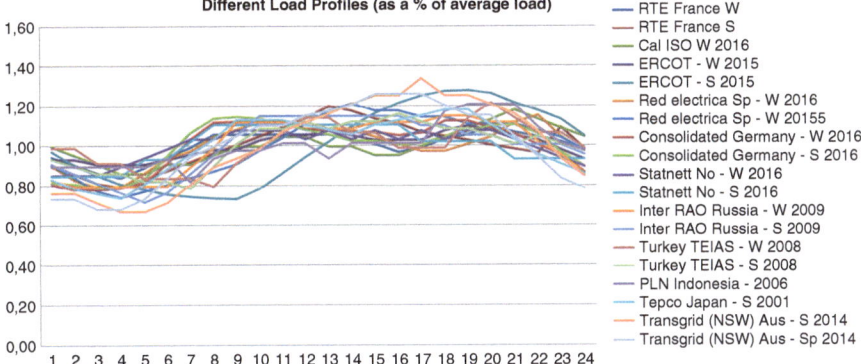

Fig. 4.15 Different load profiles (Amprion 2016; California ISO 2016; ERCOT 2016; InterRAO 2009; PLN 2006; Red Electrica 2016; RTE 2016; Statnett 2016; TEIAS 2008; Tennet 2016; Tepco 2001; Transgrid 2014; Transnet BW 2016; 50 Hertz 2016)

Load is fairly stable across the day, with differences in some regions where it actually drops around midday. Finally, a peak in demand occurs towards the end of the day, between 6 pm and 8 pm, as people get back home (Amprion 2016; California ISO 2016; ERCOT 2016; InterRAO 2009; PLN 2006; Red Electrica 2016; RTE 2016; Statnett 2016; TEIAS 2008; Tennet 2016; Tepco 2001; Transgrid 2014; Transnet BW 2016; 50 Hertz 2016). The load profiles show that Texas can reach a peak which corresponds to 1.3 times the average load (30% more than average at peak) in summer. Transgrid (New South Wales, Australia) reached up to 1.34 in summer 2014.

The peak level can also be compared to the minimum load requirement to estimate the actual ramp ups and downs needed during the day. In summer it goes up to 1.2 in California, 1.5 in Germany and France, 1.6 in Spain, 1.8 in Texas, and up to 2 in Australia. This means that, in Australia for instance, power demand doubles over the day in summer. Now, renewable energies tend to increase this already natural trend. Wind can blow at any time of the day, and the sun only shines during daytime. The consumption peak generally happens toward the end of the day, when solar power output is already in decline. The decline of solar power happens at the very same time as the actual increase in load power consumption, leading to an accelerated ramp up of conventional power capacity on the grid. Similarly, when load is dropping when renewable energies (wind and solar) are operating, at the beginning of the day for instance, there is an accelerated ramp down of the power supply from conventional sources. Since variable renewable power cannot be regulated, factoring renewable power into the load profile enables to understand the actual requirements from conventional power generation. As already explained, the net load curve is calculated by subtracting the power generation output from wind and solar energy from the actual load demand profile. The example in Fig. 4.16 shows the load curve in Germany in summer 2016 (Amprion 2016; Tennet 2016; Transnet BW 2016; 50 Hertz 2016) and the net load curve after accounting for the

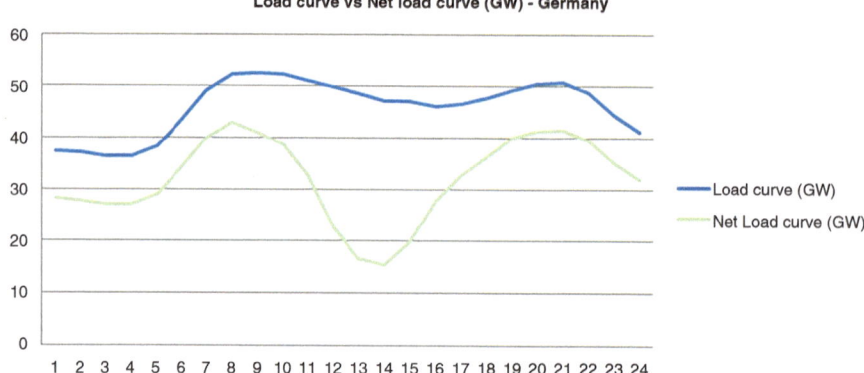

Fig. 4.16 Load curve vs. net load curve—Germany (Amprion 2016; Tennet 2016; Transnet BW 2016; 50 Hertz 2016)

impact of variable renewable energies (wind and solar). The impact of wind can be witnessed throughout the day, while the impact of solar leads to a further drop of the net load at midday. In this example, the ramp up in the morning is similar between the two curves, around 16 GW over 4 h. In the evening, the ramp up of the load curve is only 5 GW, while the net load curve shows a ramp up of 26 GW over 6 h. The impact of the penetration of a high share of renewable energies, in particular solar, has thus an important impact on how to meet the peak demand in the evening. As solar output decreases in the later part of the day, the ramp up of conventional power generation accelerates in order to meet the peak demand.

The actual ramp up at peak time can be retrieved for each region by plotting both the load curve and the net load curve and measuring the actual ramp up volume (GW) over a period of 4 h. The 2030 estimates can also be retrieved by estimating the evolution of the load curve (it is assumed here the shape remains unchanged) and recalculating the net load curve using the forecasts of renewable penetration over the grid by 2030, as already explained. The ramp up volume (GW) is plotted on Fig. 4.17 in today's situation, and in 2030 according to the selected scenario of renewable penetration. In all regions studied the volume of ramp up rises significantly. It increases notably in California and Germany (and to a lesser extent in Texas) where it respectively reaches 40% and 50% of the maximum load by 2030, compared to an average current level around 20%. In other regions, the impact is less significant.

To summarize, the peak load is the result of power consumption usages. They vary from one region to another because of the mix between the different sectors (residential, commercial buildings, industries and transportation), as well as because of the specific usages with each sector. Globally, peak consumption is generally identified at the end of the day, when people leave the office and return home. This peak traditionally occurs between 6 pm and 8 pm in the evening. The size of this peak also varies between seasons because usages vary across seasons. Electricity markets are prepared for these peaks, with specific peak power generation capacity

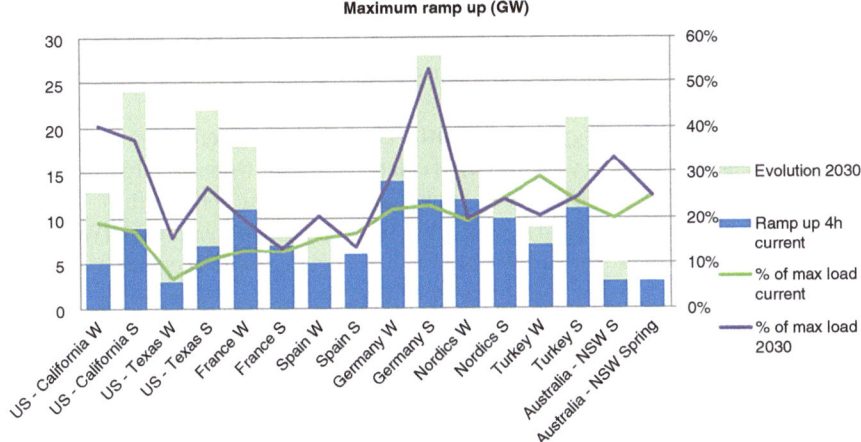

Fig. 4.17 Maximum ramp up (Source: Author's own calculation)

turned on at peak times in order to meet the increased demand. The increase in the penetration of renewable energies, in particular solar, tends to decrease the need for conventional generation during midday. This leads to an increased size of the ramp up, in particular when solar energy output drops back to zero. One of the key effects of renewable penetration is thus the increased need for peak generation capabilities. In a number of regions, such as California and Germany, these capabilities are expected to increase significantly in the coming decades if no other solution steps up, such as demand shifting and energy storage. This would lead to an increase in overall costs, especially during ramp ups. These units would indeed only be able to amortize their costs during this short period of time.

4.2.3 Facing Increased Uncertainty: Balancing Services

Supply and demand of power need to be balanced at all times, as electricity today cannot be efficiently stored. There are typically three levels for a grid to respond to a variation in the balance between supply and demand. The first level, called primary control, corresponds to the "spinning" reserve available on conventional rotating machines. Each generator can adjust its production upwards or downwards in order to adjust to the demand. This primary control can usually be done in a matter of seconds. Then, secondary control corresponds to the call for balancing services which can be activated in a matter of minutes. This can correspond to adjustments of production on current generators operating on the network, or actual ramp up of units such as hydropower plants, open cycle gas turbines or diesel generators. This typically complements the primary control and helps bring back frequency and voltage back to nominal levels. Finally, tertiary control aims to bring back the amount of reserves and the production planning in line with the expected upcoming contingencies. Balancing services typically correspond to the secondary and tertiary controls and are typically designed to adjust to demand prediction issues or the loss of a generation unit.

Variable renewable energy output is difficult to predict, therefore the use of these services tends to increase. Typically, wind can be predicted with a +/−15% accuracy 1 day ahead (© OECD/IEA, Power Transition 2012) and solar with a +/−12% accuracy (de Felice et al. 2014). According to the German institute (DIW 2011), forecasts on wind in Germany are becoming more accurate, with accuracy as good as +/−7%. Nevertheless, the predictability of demand is much more accurate, typically between 2% and 4% (Taylor et al. 2006). There has been a number of studies to show that the integration of renewable energies into the grid at various locations has a positive impact overall on the predictability of the supply, since the errors tend to compensate one with another. This would lead to a reduction of the error percentage for the supply forecasts. This being said, the more renewable energies on the grid, the higher the volume of possible errors, and thus the volume of required balancing services (© OECD/IEA, Costs 2015). Figure 4.18 shows the actual variation of the "net" load curve in Germany based on a +/−10% error for the renewable output forecast. The variation between the two is the actual volume of generation which needs to be held in reserve in order to compensate for this uncertainty.

Renewable energies thus lead to an increase in the need for balancing services. These needs can be estimated based on the actual volume of variable renewable energies on the grid. Using the estimates from Fraunhofer (2016), the production output of renewable energy can be derived at one point in time for a given region. The one-day-ahead predictability issue can be applied to this volume to yield the actual additional requirement for balancing capacity. Then, using the forecasted evolution of demand and of renewable penetration, the same value can be estimated for the period to come, against a number of scenarios. Assuming a 10% error on renewable supply forecasts on average, Fig. 4.19 shows the actual impact of further

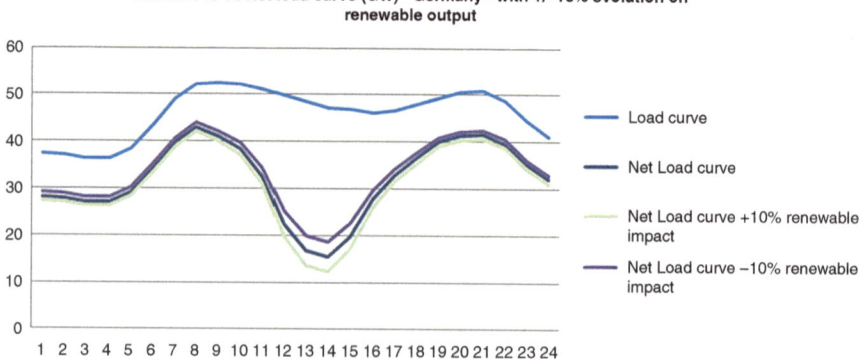

Fig. 4.18 Load curve vs. net load curve in Germany with +/−10% evolution on renewable output (Source: Author's own calculation)

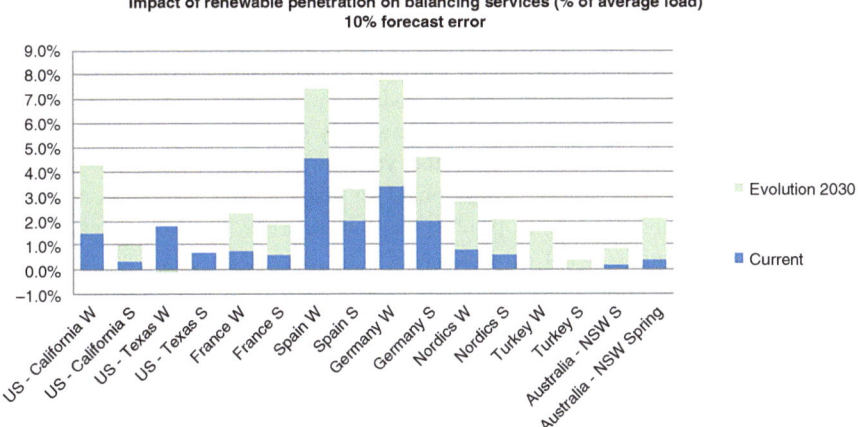

Fig. 4.19 Impact of renewable penetration on balancing services (Source: Author's own calculation)

renewable penetration on the required balancing reserve to be available as a percentage of the average load. If we bear in mind that the actual demand predictability shows errors in the range of 2–4% maximum, the actual lack of predictability of the balance between supply and demand due to renewable energies strongly increases the need for balancing services. This is already the case in Spain and Germany, with respectively 4.5% and 3.4% of the average load held as a balancing reserve to cope with renewable penetration forecasting issues. These ratios are lower in other regions of the world but expected to rise strongly in the coming decade with the further penetration of renewable energies. They would increase up to 2% in most regions of the world, and reach as high as 4.3% in California, 7.4% in Spain and 7.8% in Germany! These ratios are obviously theoretical as they do not differentiate wind and solar production forecasting and do not take into consideration the possible improvements in forecasting which could be achieved in the coming years.

Higher renewable penetration into the grid clearly leads to a strong increase in the need for balancing services. Hirth (2012) conducted several studies to evaluate the impact of the penetration of wind in electricity systems. He concluded that balancing costs are expected to increase everywhere. While the traditional (without renewable) volume of balancing costs is estimated to be between 3% and 4% of the total value of electricity, the higher penetration of renewable energy into the grid leads these costs to increase. They could as much as double in certain regions (the Nordics) and increase by around 50% in other regions (the United States). They remain low in comparison to the costs associated with the drop in the capacity factor of conventional generation. However, they still add up a few percent of the final value of electricity. In 2014, they represented one billion British pounds for the national operator in the United Kingdom, having risen by almost 40% in the preceding 10 years (NAO 2014). In the Nordics, Statnett reported its overall system operation costs to have been multiplied by almost five in 10 years (Statnett/System 2015); the penetration of renewable energies accounted for part of the overall increase.

One of the ways to compensate for this lack of predictability and its associated costs is to shorten the forecast period. Today, most power markets operate on a day-ahead basis, which means that power supply commitments are taken 1 day in advance. The lack of accuracy of the variable renewable forecasts leads to volumes of balancing capacity being upheld so that the operator can compensate for variations in the renewable production output. The "Deutsches Institut für Wirtschaftsforschung" (2011) showed clearly the advantage of create an "intra-day" market, one as continuous and liquid as possible, in order to favor last-minute adjustments from the market, therefore reducing the actual amount of the planned reserve. The development of such markets remains however to be organized in many countries. They also represent an important opportunity for operators to optimize their revenue in the context of growing market complexity.

4.2.4   Real-Time Frequency Control Issues: Ancillary Services

Traditional electric systems operate at a fixed frequency, usually 50 Hz or 60 Hz, depending on the region of the world. All generators operate at this frequency, and connected equipment and appliances require this exact frequency to function properly. Any important variation of the frequency can damage appliances connected to the grid, as well as create network stability issues and massive energy distribution disruption. Most conventional generators have rotors powered by a turbine, with these rotors synchronized to the frequency of the network output. They thus rotate at a certain speed which is imposed by the network. The energy that the turbine transmits to the rotor enables it to modulate the power output to the grid. The rotational speed remains set at the nominal frequency of the grid. When a major imbalance occurs on the network, such as a generator brutal decoupling, or a demand surge, the different generators connected on the network need to compensate. When confronted with a sudden request for increase in output, for a given input of the turbine, the speed of generators tends to be affected. They react like a rotating machine facing a higher "resistance". Generators, because of their inertia, resist this higher resistance and slow down. As they slow down, the frequency drops. Then, operating controls help adjust the turbine inputs to bring back the system to normal. This is the "inertia" principle which defines the stability of grid network operations.

Now, renewable energies such as wind and solar are not synchronized to the grid. Both produce current which is then turned into 50/60 Hz AC current through electronic devices such as inverters. There is no "inertia" in electronics. Wind and solar energies thus do not contribute to the global "inertia" of the grid. This remains of little importance as long as the ratio of conventional energy in the grid remains high. However, if the instantaneous contribution of conventional generation to the grid drops below 40%, the corresponding lack of inertia on the grid may lead to important network stability issues (MacGill and Riesz 2012; NREL/Wind integration 2014).

The evolution of the minimum ratio of synchronous machines can be estimated based on the current volume of renewable capacities on the network as well as the evolution expected in the years to come. The minimum load balancing (minimum volume of conventional generation required) is divided by the average load to give the minimum rate of synchronous machines on the grid. Again, this is a theoretical assumption, but it gives an indication of the extent of the issue.

Figure 4.20 shows a decrease of the rate of synchronous generators in all regions because of the higher penetration of renewable energies. Many grids will essentially reach, at least at some point during the day, less than 40% of penetration for conventional sources (60% of renewable generation). Spain and Germany are already in this situation, but the fact they are interconnected to the European network makes this ratio artificial, as other countries of Europe can contribute to the overall inertia of the network. Now, as all countries in Europe move to renewable energies, the global inertia of the interconnected network could be put at risk. As well, inertia and frequency control can only be achieved when networks between regions are made through AC lines. High-voltage DC connections (HVDC lines) cannot help as they follow the same principle as inverters in wind farms and isolate networks. By 2030, California shall also reach a low point.

The forecasts all lead to a drop in the network's inertia. There are only a few solutions to increase stability in such a context (Tielens and Van Hertem 2012). The wind or the solar farm operation can be regulated to either create a "reserve" through artificial deloading of the farm, or to electronically command a greater output by using the kinetic energy of the rotor of a wind turbine. These solutions lead

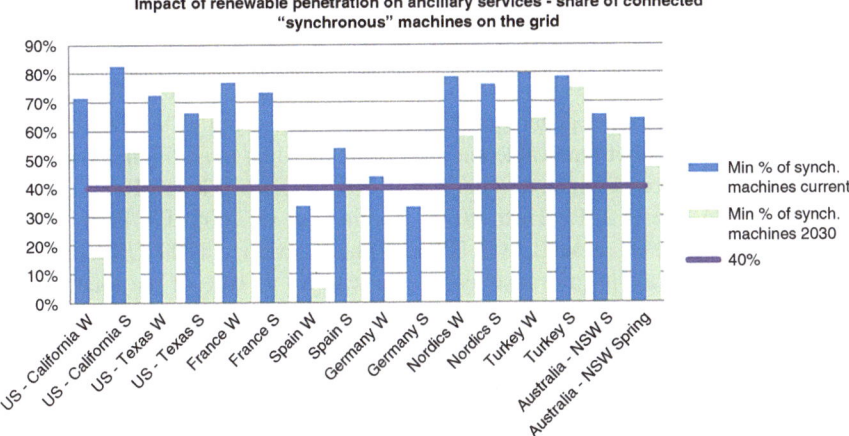

Fig. 4.20 Impact of renewable penetration on ancillary services (Source: Author's own calculation)

renewable farms to operate at a lower efficiency point than their optimal set point. The latter solution cannot efficiently support the grid frequency for a long period of time, leading invariably to the wind farm output dropping after some time as the "input" to the turbine (the wind in this case) cannot be artificially adjusted. Another solution which could partially answer the problem consists of energy storage systems which could be coupled to renewable farms and be used as an immediate power supply support (and therefore grid frequency support) in case of a sudden imbalance on the network. Finally, coordinated demand side management also presents a way to recover balance between the supply and demand sides. We will explore further some of these solutions in the next chapter.

4.2.5   Additional Grid Costs

Large renewable utility-scale farms are often located in areas where either sun or wind lead to efficient capacity factors. In addition, these farms are not necessarily close to large urban centers. This leads to further losses in the transportation of energy from the source of supply to the consumption point. Typically, networks are today optimized to supply energy from generation sources (usually close to urban centers) to consumption areas. The increased penetration of renewable energies is changing the architecture principles that guide grid network design. As most renewable energies are located in distribution networks, they lead to zones with strong underutilization of network capabilities and, contrarily, network areas with congestion issues. Indeed, areas of the network designed to supply important amount of energies can become underutilized if locally renewable energies are installed (this could be particularly demonstrated with the development of self-generation). On the contrary, the installation of renewable energy on parts of the grid with no generation and only small volumes of distributed consumption might find these parts congested by too high a flow of energy. Finally, the high penetration of renewable energy is leading to increased requirements for balancing and ancillary services. Interconnection between regions is a good way to stabilize the network and facilitate the participation of additional generation sources to the balancing of the network. The increased penetration of renewable energies is thus leading to additional grid costs. These costs have been estimated to range between 2 USD/MWh and 10 USD/MWh (© OECD/IEA, Costs 2015; Hirth 2015; Holttinen et al. 2008; PV Parity 2013). With a retail price of around 100 USD/MWh on average (200 USD/MWh in Europe), this increase in grid costs corresponds to up to 10% of the total cost of electricity. In addition, these costs could even be higher since a relatively low penetration rate of below 50% in general (except in the case of Ireland) is assumed here, and the rate tends to grow with the increased penetration of renewable energy.

Going further, the main equation beyond the costs of the impact of renewable energies on the grid is as well the profitability profile of the investments realized. It is important to understand that a reasonable rate of return is expected from these investments. Since retail prices are set by regulations, the profitability of these investments is globally determined by the regulator. If the assets invested are not properly utilized, the cost of transmission and distribution becomes economically not viable. In addition, grid electricity is already competing with distributed rooftop solar energy. In many countries, grid parity has already been (or is close to be) reached. Unwise and unprofitable investments thus lead to extra grid operation costs, which lead to a further lack of competitiveness of grid distributed energy as opposed to localized distributed energy production.

Currently, assets are already underutilized. There is no easily available public data to evaluate the overall assets utilization of transmission and distribution infrastructure. More importantly, assets utilization is difficult to average overall since they vary from one part of the infrastructure to another. Some parts of the network may indeed be constantly under stress, especially in the case of remote locations, while others may actually be less utilized in general, notably in regions with a high density of networks distribution (close to urban centers for instance). Global averages can be calculated to estimate the overall asset utilization of a given network. Following the actual load curve of a region, it can be assumed that the network is designed to support the maximum amount of load which is required at a point in time. This is obviously a raw approximate and should thus be considered carefully. From this can be derived the average utilization rate of the grid. The analysis of a number of utilities yields an average utilization rate of around 70% of grid assets (Fig. 4.21). This varies significantly across seasons. It can be as low as 40% in winter for California and reach 80% in summer. It can get as high as 90% in France, Germany or the Nordics countries in winter, go down to 50% in France and the Nordics in summer, and 70% in Germany. Spain shows a different trend with very similar ratios between summer and winter of around 80%.

Another way to look at asset utilization is to evaluate how much of the time assets are underutilized. Three measures have been made in Fig. 4.22 to evaluate from the load curve how much percentage of the time assets were operating below 50% of the maximum load, 60% of the maximum load, or 70% of the maximum load. Important

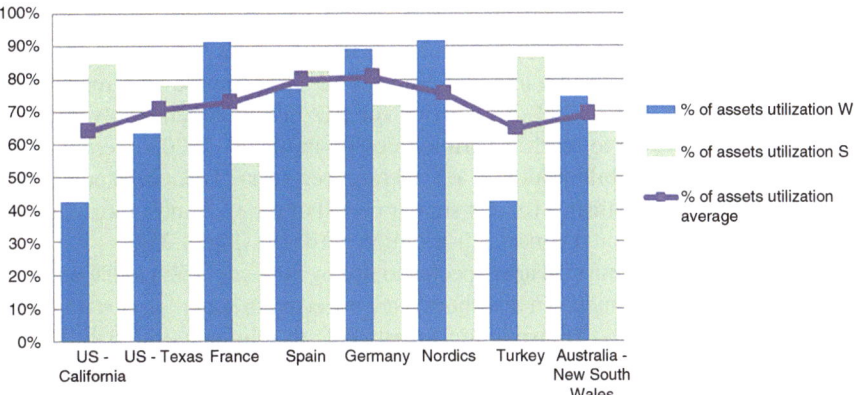

Fig. 4.21 T&D asset utilization (Source: Author's own calculation)

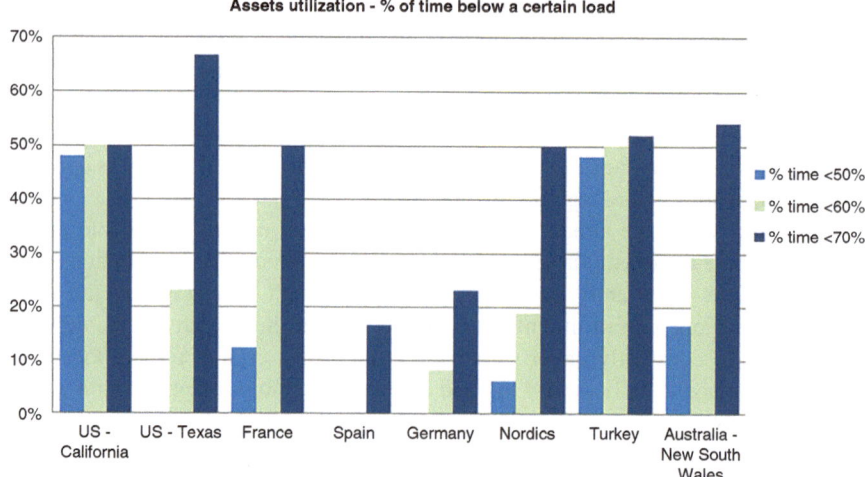

Fig. 4.22 T&D asset utilization—percentage of time below a certain load (Source: Author's own calculation)

differences exist across regions. California and Turkey's assets are massively underutilized, with their grid operating below 50% of the maximum load almost 50% of the time. The situation is reversed for Spain and Germany, which operate above 70% of their maximum load around 80% of the time. Texas and the Nordics almost never operate below 50% of their maximum capacity, they however operate more than 50% of the time below 70% of their capacity. This means they generally operate between 50% and 70%. France operates only 10% of the time below 50%, but half of the time above 70% of the maximum load, a sign of great variation of load. Australia has a similar profile, operating below 50% of the load close to 20% of the time, but operating above 70% of the load half of the time.

The utilization of assets thus varies significantly across regions. The profitability of certain types of investments is thus questionable in many parts of the world, and each project is obviously checked carefully. Grid investments are expected nevertheless to increase significantly in the coming period. These investments represented 2.9 trillion dollars in the 2000–2012 period (or 212 billion dollars on average per year) and are expected to reach 5.8 trillion dollars in the 2012–2030 period in the CPS scenario (or 323 billion dollars on average per year). This corresponds to a growth of 50% in investments (over a similar period of time). Almost three quarters of these investments would concern distribution networks (Fig. 4.23).

Investments in transmission are expected to rise by 80% compared to the previous period, while the investments in distribution are expected to "only" rise by 45% (but the baseline is stronger). Transmission investments to reinforce grid stability and enable balancing services are thus expected. On the other side, distribution investments are also expected to rise significantly to reinforce the grid, especially in areas where renewable energies are installed.

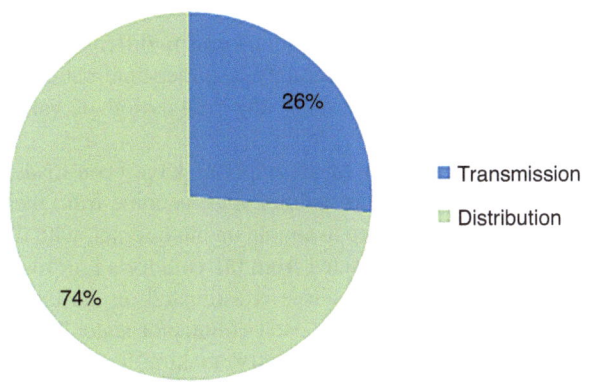

Fig. 4.23 Split of investments in the grid (© OECD/IEA, WEIO 2014)

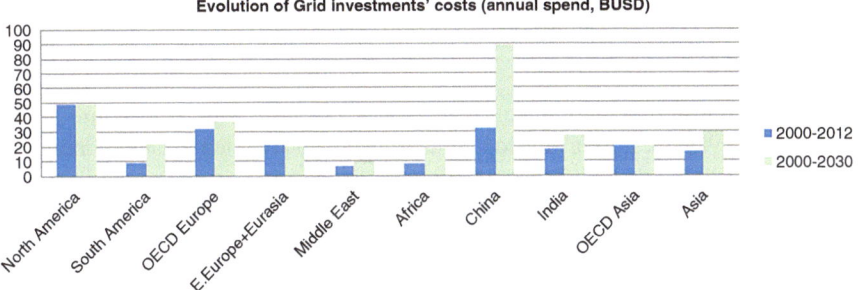

Fig. 4.24 Evolution of grid investments' costs (© OECD/IEA, WEIO 2014)

These investments are expected to rise everywhere but they will grow more substantially in new economies such as South America, India, Asia, Africa and, more importantly, China. China only represented 15% of the worldwide investments in the 2000–2012 period, and is expected to represent 28% of the worldwide investments in the coming period. Investments would almost triple there (Fig. 4.24).

Ahead of those massive investments, the key question is whether or not they will be profitable and if financing will be available. Since the transmission and distribution of electricity is a regulated activity, these investments are amortized on the cost of electricity supplied to the consumers. On one hand, localized distributed energy could partially disrupt the model by offering to consumers cheaper sources of energy which would not require (or perhaps partially) the use of the grid. On the other hand, investors may not be able to secure such massive financing, despite the theoretical guaranteed rate of return. Evolutions in regulation and in the overall economic situation will tell if and how these investments will happen.

4.2.6   *A Regional Perspective*

The transition to a new grid can take different forms in different countries. It is thus important to analyze each of them separately to understand better the issues today and the challenges ahead with regards to the penetration of variable renewable energies into the grid.

As already presented, a number of assumptions have been made in the figures displayed below. The load curves in various seasons come from the official publications of the various utilities and are available on the Internet, with the exception of Germany, where the data is consolidated from the country's four main transmission operators. The 2030 forecasted load curve is estimated on the base of the overall evolution of the power consumption in each country or state. The actual share of renewable energy is available for each country, making it possible to estimate the net load curve using assumptions derived from Fraunhofer (2016) and explained above. The forecasted rate of penetration is estimated using a number of forecasts (as described above).

4.2.6.1   France

Electricity consumption in France has a strong correlation with temperature. In the 1980s and 1990s, the strong development of electric heating led to an important gap of consumption between winter and summer season. The average load in winter ranges around 75 GW and around 45 GW in summer, a 30% gap. The load curve in winter is fairly stable during the day, with a peak in the evening which can reach up to 100 GW on some specific days, and a lower level at night. In summer, the load is less constant, with a peak around midday averaging around 55 GW (Fig. 4.25).

The load is estimated here to grow by around 5% in the coming decades, which corresponds to an annual growth rate of around 0.3% (RTE/Demand 2015). RTE

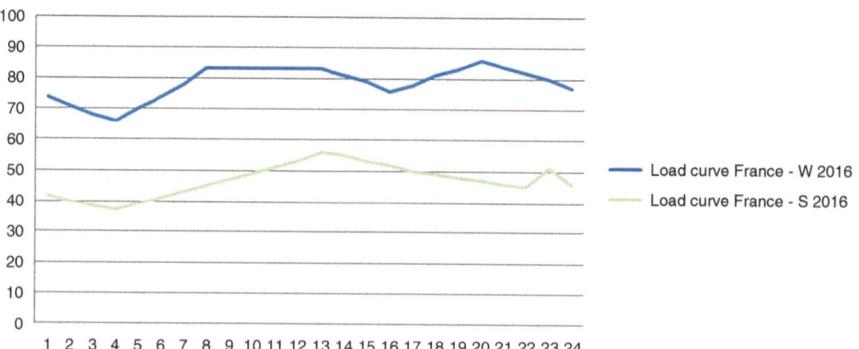

Fig. 4.25 Load curve 2016—France (RTE 2016)

(2015) has indeed projected an average growth till 2020 of no more than 0.3% (reference scenario). This forecast was expanded to 2030.

There is today around 10 GW of wind and 6 GW of photovoltaic solar connected to the grid (RTE/Renewables 2015), which add up to around 12% of the total capacity available (around 130 GW). This leads to a net load curve which is slightly different from the actual load curve, since variable energies operate at a low capacity factor. As a reminder, it has been assumed here that wind operates with an average of 26% in summer and 56% in winter, while solar operates only between 9 am and 6 pm at up to 33% of its capacity in winter and 59% in summer (Fraunhofer/Renewable 2016). Because of the very low base of photovoltaic solar, the share of renewable generation is generally relatively stable, therefore the net load curves look very much alike the actual load curves. Minimum load balancing varies between 35 GW and 60 GW, depending on the season. The volume of generation expected to ramp up in less than 4 h is limited to around 10 GW as the load curves are fairly stable. There is a high level of synchronous machines on the network (Fig. 4.26).

By 2030, the share of renewable energies is expected to increase strongly in France. The selected scenario estimates that France could have up to 35 GW of wind and 30 GW of solar. The trends remain the same, with a greater gap between the actual load curve and the net load curve.

Globally, the minimum load balancing is marginally impacted despite the increase in renewable generation. The impact on base load generation is therefore limited in France. The largest impact occurs in winter, when net load requirements reduce by around 10 GW. It however remains higher than summer in any case, with consequently a limited impact on the base load supply. Some existing generation capacity could however not be able to operate continuously in winter anymore. This could lead to putting in place some capacity mechanisms to secure the profitability of these units (if they are found to be useful on the grid) or to a certain level of restructuring. The maximum load is also marginally impacted. The volume of energy to ramp up at peak is however increasing by up to 7 GW. This additional load should

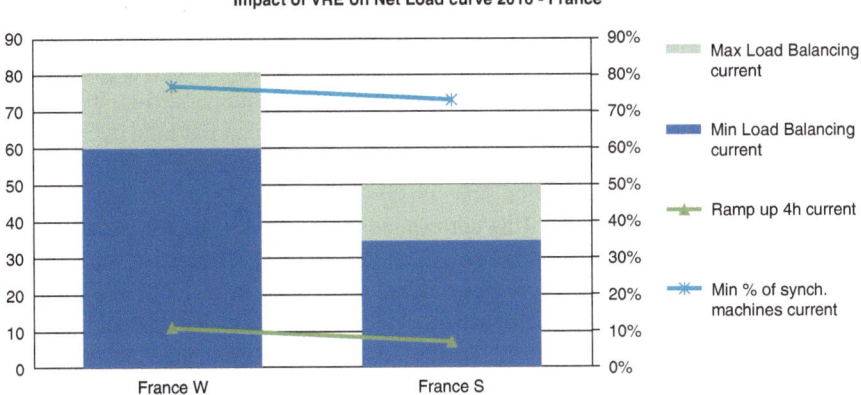

Fig. 4.26 Net load curve 2016—France (Source: Author's own calculation)

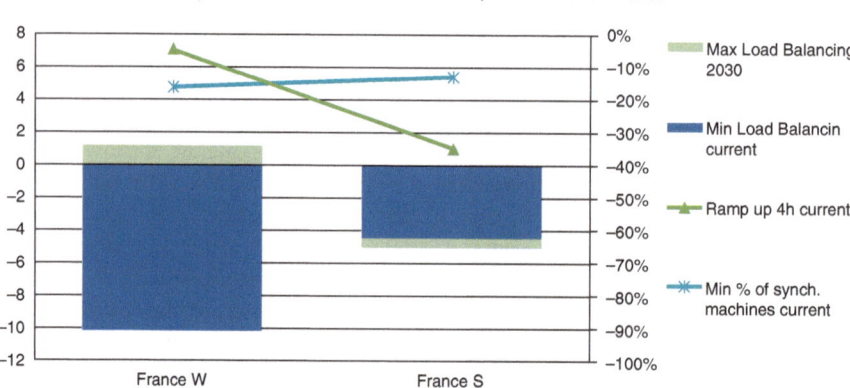

Fig. 4.27 Net load curve 2030—France (Source: Author's own calculation)

thus be supplied by a mix of mid-peak and peak units. The high penetration of renewable leads to no more than 2% additional forecast error on the average load in the worst case scenario. In all scenarios, the share of synchronous machines remains well above 60%, leading to no issues of network stability. France is thus not expected to have any strong disruption to its power generation footprint in the coming decades (Fig. 4.27).

4.2.6.2   Germany

Germany's load curve also depends a lot on seasons and weather conditions, although to a lesser extent than France. The load goes up to 65 GW on average in winter and slightly above 50 GW in summer. The difference is however much less pronounced than in the case of France. The load curve typically ramps up in the morning and remains flat throughout the day, with a small peak in the morning and another one in the evening (Fig. 4.28).

The load is estimated to remain flat in the coming decades. Five-year projections indeed show a flat to downward trend (Renewables International 2014). It has thus been assumed that the increase in consumption would remain flat until 2030.

Germany has developed renewable energies much faster than France. There are today more than 35 GW of wind and 38 GW of photovoltaic solar on the grid (Fraunhofer/Renewable 2016). This corresponds to around 40% of the total power generation capacity connected to the grid. This leads to very specific net load curves, with a much smaller demand for conventional generation in general, and a gap at midday when solar is at work, in particular in summer. The share of mid-peak to peak load is thus important, up to 30 GW in summer, with ramp up capacities ranging around 15 GW. The rate of synchronous machines is already lower than 40% in summer (Fig. 4.29).

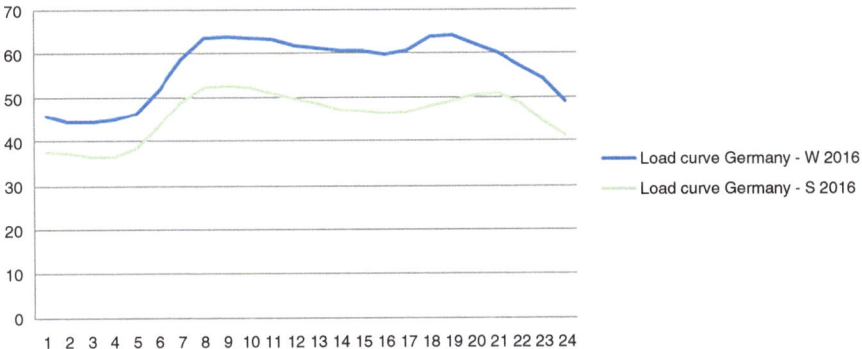

Fig. 4.28 Load curve 2016—Germany (Amprion 2016; Tennet 2016; Transnet BW 2016; 50 Hertz 2016)

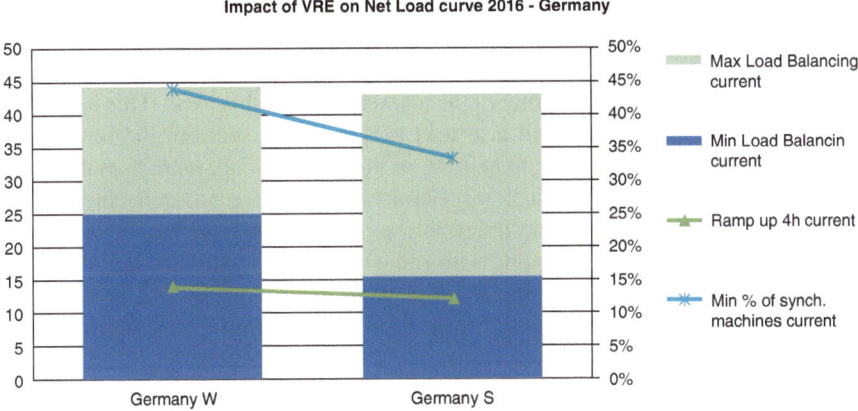

Fig. 4.29 Net load curve 2016—Germany (Source: Author's own calculation)

By 2030, the assumptions taken lead to a continued growth of the renewable penetration into the grid, with up to 80 GW of wind and 75 GW of solar. This high penetration leads to a complete change of paradigm for electricity markets. No generation capacity will be required continuously on the grid, since the load required from conventional generation will be zeroed a significant part of the time in both winter and summer. As a result, the 15 GW of base load generation which today can operate continuously on the grid will face a significant capacity factor issue.

As a consequence, most of the base load generation will end up being restructured to the benefit of mid-peak and peak load generators. The peak load is also expected to go down, from around 43 GW today to between 20 GW and 30 GW, depending on the season. The volume of conventional generation required to ramp up in 4 h is however expected to double, up to nearly 30 GW in summer, leading to a significant

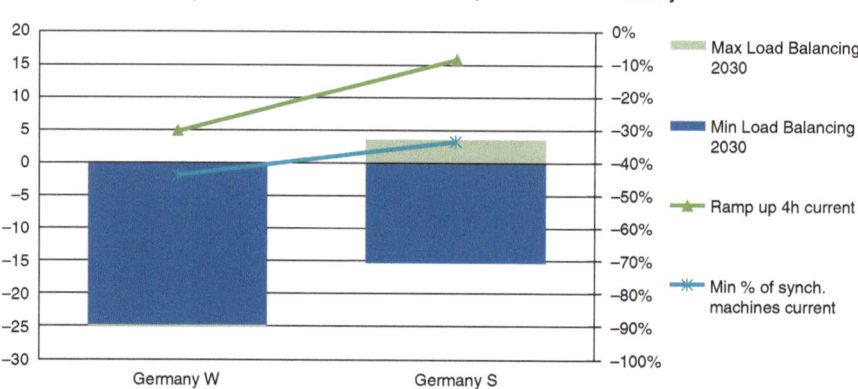

Fig. 4.30 Net load curve 2030—Germany (Source: Author's own calculation)

need for peak load units. In essence, the share of base load is expected to be substituted by mid-peak and peak units such as gas turbines or alternative technologies. A strong change in the power generation mix is thus expected to happen in Germany. The current restructuring of the market confirms this trend. This important share of renewable energies will also lead to a strong increase in the forecasting errors on the power generation, up to 7.8% of the average load, more than the triple that of France. It is thus critical for Germany to develop tools (technical and economic) to better forecast the renewable production output. Finally, since the German grid could end up operating with zero conventional generation, the stability of the grid could be put in question, although the network remains interconnected to the rest of Europe. Germany will thus be one of the "smart grid" laboratories of the coming decade (Fig. 4.30).

4.2.6.3 Spain

Spain presents a very different load curve. Unlike in France and Germany, its power consumption is higher in summer than in winter. This is essentially related to local weather conditions, with extremely hot summers and soft winters. As a result, air conditioning is favored over heating. The load curves of Spain show this difference. If the load goes up to 35 GW in summer, at midday, it generally ranges around 30 GW in winter. The peak also generally occurs around midday, except in winter, when it occurs in the evening, later than in France and Germany, because of the characteristics of Spanish lifestyle habits (Fig. 4.31).

The load is estimated to grow by around 20% in the coming decades, which corresponds to an annual growth rate of around 1.3%, close to the average projected evolution of consumption in Europe across different scenarios.

Spain has long been developing renewable energies. There are today more than 23 GW of wind power in Spain, and around 4 GW of photovoltaic solar, together

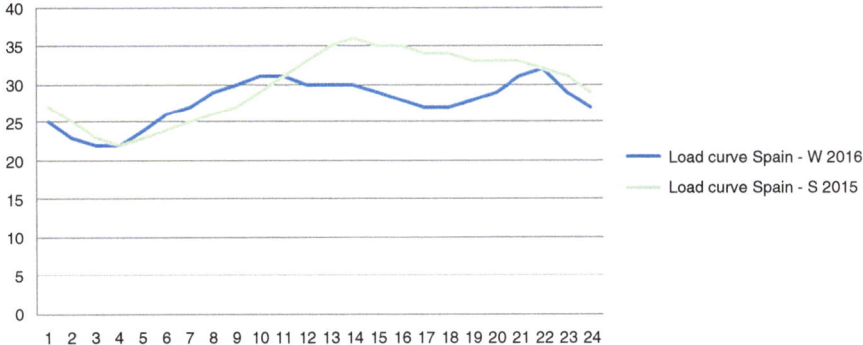

Fig. 4.31 Load curve 2016—Spain (Red Electrica 2016)

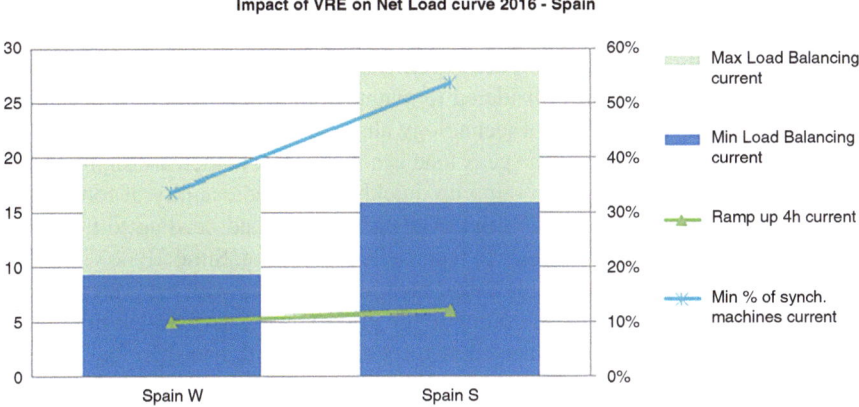

Fig. 4.32 Net load curve 2016—Spain (Source: Author's own calculation)

around 30% of the total capacity installed (Red Electrica/Renewable 2013). The net load curve is thus strongly impacted by the share of renewable energies in the grid, although solar is not extremely significant. Because of the wind capacity installed, the minimum load balancing can today get as low as 10 GW. The volume of mid-peak to peak load is significant, around 12 GW, with 5 GW required to ramp up in 4 h. Already Spain can experience situations with a ratio of synchronous machines below 40% for part of the day, notably in winter (Fig. 4.32).

The selected scenario for 2030 leads to 45 GW of wind and 16 GW of solar. Because of the rather low penetration of solar (the forecast could end up being wrong as Spain is a good candidate for solar in terms of irradiation level), the main impact on the net load curve comes from the strong increase in wind power on the grid. While base load averages 10 GW in winter and 15 GW in summer today, it would go down to 2 GW in winter and remain globally stable around 15 GW in summer.

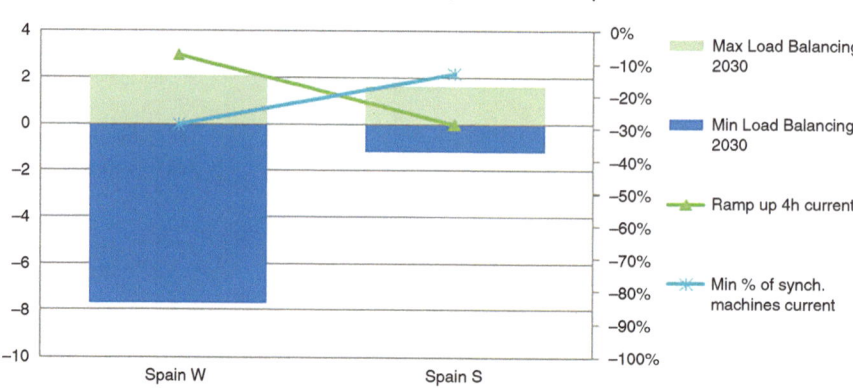

Fig. 4.33 Net load curve 2030—Spain (Source: Author's own calculation)

   Base load generation would not be able to operate continuously throughout the year anymore, though it could reach significant output during summer. Capacity markets could end up being considered to maintain the actual volume of base load units operating, in the absence of technology alternative, even though the economic equation will be undermined. The peak load is not expected to increase significantly, nor the volume of generation to ramp up quickly. The predictability of forecasts is expected to yield around 7% of errors over the average load, leading to flexibility mechanisms and to strong attention to renewable forecasting. Similarly to Germany, the high penetration of renewable could also lead to times with very low to zero contribution of conventional generation to the grid, thus to possible stability issues, although Spain is also interconnected to the European grid, and therefore benefits from it. As a consequence of this, Spain is also expected to be one of the pioneers of "Smart Grid" development (Fig. 4.33).

4.2.6.4   The United States—California

The California load curve depends highly on the season. The load barely tops 25 GW in winter while it doubles in summer to reach around 50 GW. The load curve is flat during the day, with a slight drop at night and a peak around the end of the day. The peak is more pronounced in summer with a load which can reach 52 GW for an average load in the day of around 45 GW (Fig. 4.34).

   The load in California is estimated to grow 1.4% on average, leading to around 24% of electricity consumption growth by 2030. This is estimated by using the actual forecasted growth of consumption of North America (© OECD/IEA, WEO 2014) applied to the actual load of California.

   California has only 6 GW of wind power and 5 GW of solar power installed today (California ISO/Renewable 2016). This corresponds to less than 13% of the total generation capacity installed. The net load curve is thus currently not strongly

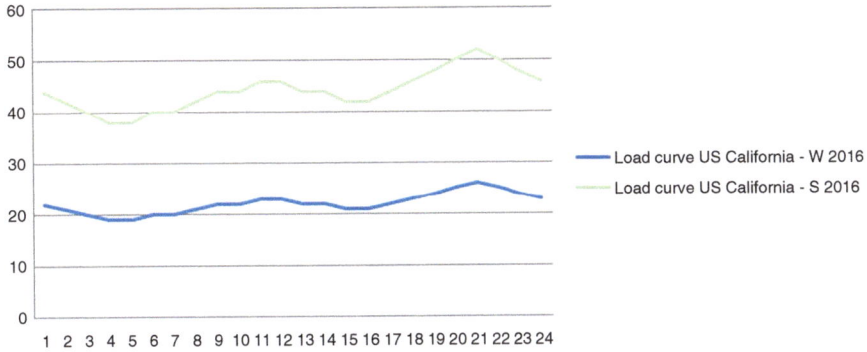

Fig. 4.34 Load curve 2016—California (California ISO 2016)

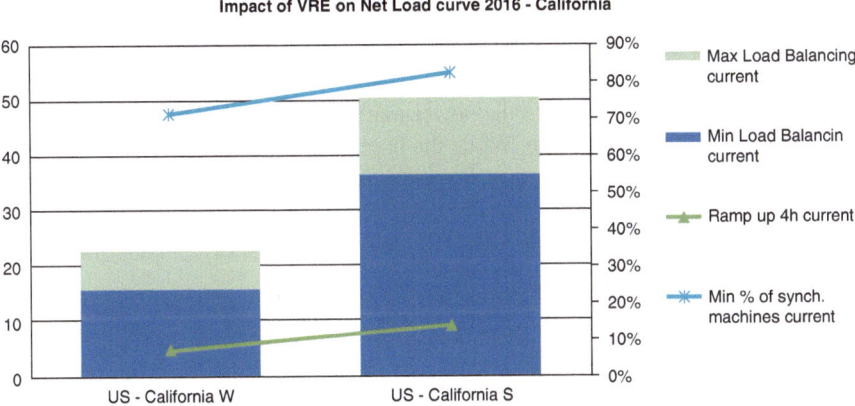

Fig. 4.35 Net load curve 2016—California (Source: Author's own calculation)

impacted by renewable energies. The impact of solar starts however to be witnessed around midday, but the gap with the actual load curve remains minimal. Minimum load balancing varies between around 15 GW and 35 GW, depending on the season. The ramp up capacity remains low at less than 10 GW, and the share of synchronous machines is above 70% (Fig. 4.35).

The rate of penetration of renewable energies is expected to accelerate in the coming years. The selected scenario yields 21 GW for wind and 34 GW for solar by 2030.

The high share of solar in the overall generation mix leads to a complete change of paradigm for the power industry in California. While the load in winter is today fairly stable, a significant drop will occur by 2030 because of solar power, and the load required from conventional generation is expected to drop down to 4 GW around midday. There would thus be virtually no more base load in California by this time. The picture is slightly different in summer, when the load is much higher,

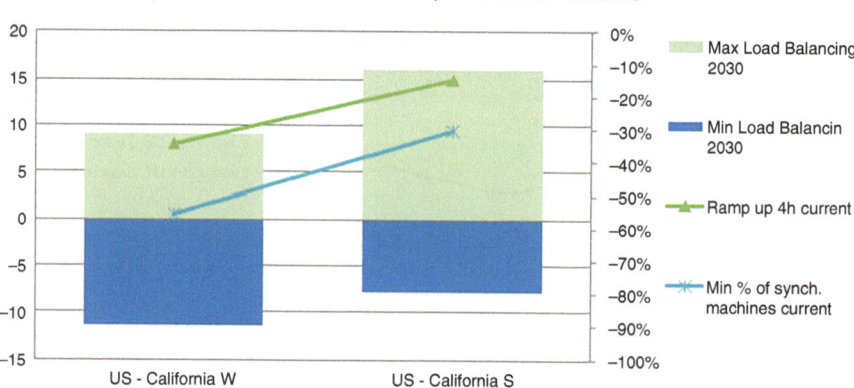

Fig. 4.36 Net load curve 2030—California (Source: Author's own calculation)

typically around 35 GW of base load at present. The net load is expected to drop down to less than 30 GW by 2030.

As a consequence, changes to the power market are expected to be significant in California in the decade to come. While the base load is ranging between 15 GW in winter and 35 GW in summer today, it is expected to drop down to 4 GW in winter and to 28 GW in summer. The restructuring of a part of the conventional generation fleet will thus be required, along with the development of new market mechanisms (capacity, etc.).

The maximum load is expected to increase significantly (15 GW) because of the forecasted growth in power consumption. Consequently, the ramp up of peak units is expected to increase, from a high single-digit figure (5 GW in winter and 9 GW in summer) to up to 14 GW. Summer will specifically lead to a stronger ramp up and therefore require additional peak units. The market is thus expected to restructure towards a lower number of base load units, and more mid-peak and peak units. Also, the high rate of renewable penetration will lead to a strong increase in forecasting errors (up to 4%), which would in turn lead to an increase in balancing services from the grid operator. Finally, California will also be subject to a strong decrease in the volume of synchronous machines connected on the network, leading to higher sensitivity to stability issues. Like Germany, California will become a perfect example of the changes at stake and upcoming challenges for grid operators (Fig. 4.36).

4.2.6.5   The United States—Texas

The Texas load curve depends highly on the season. While winters are cold, summers are extremely hot. This leads to a fairly flat winter load curve of around 40 GW, and a summer load curve which goes up continuously during the day to reach a peak of around 65 GW. Air conditioning likely plays a key role in this load curve, as temperatures regularly go above 40 °C in summer in Texas (Fig. 4.37).

The load in Texas, like for California, is estimated here to grow at an average annual rate of 1.4%, or 24% of electricity consumption growth by 2030. This is consistent with ERCOT's long-term forecast (ERCOT/Projections 2016).

Texas has virtually no photovoltaic solar, with around 1 GW only of capacity. It has however around 14 GW of wind (ERCOT/Renewable 2016). In total, variable renewable energies represent around 19% of the total capacity installed. The shape of the net load curve is thus barely impacted in our model since there is virtually no solar. The volume of conventional generation required is lower than the load curve because of the contribution of the wind capacity, with a stronger impact in winter than in summer. Minimum load balancing represents around 30 GW, and mid-peak to peak units another 30 GW (in summer), but less than 10 GW (in winter). Ramp up generation is limited below 7 GW and synchronous machines exceed 70% of the total (Fig. 4.38).

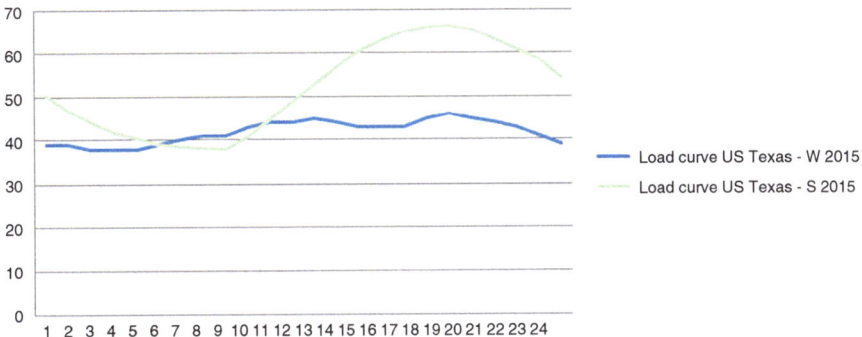

Fig. 4.37 Load curve 2016—Texas (ERCOT 2016)

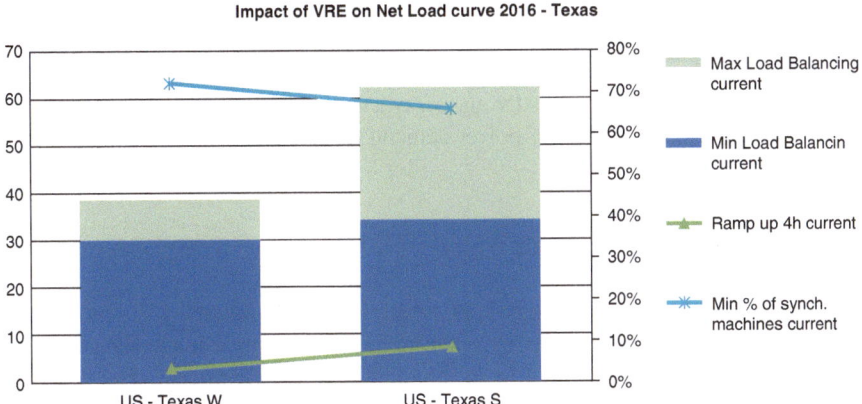

Fig. 4.38 Net load curve 2016—Texas (Source: Author's own calculation)

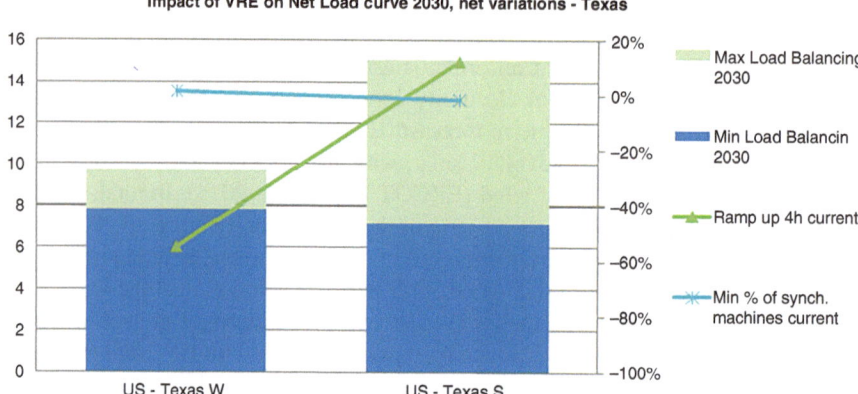

Fig. 4.39 Net load curve 2030—Texas (Source: Author's own calculation)

Following the forecasts of renewable development in North America, the rate of penetration of renewable energies in Texas is expected to increase to 16 GW for wind and 21 GW for solar by 2030. Wind was developed first because it was competitive to operate. A strong growth of solar is now expected.

The minimum load balancing is expected to increase to 38 GW in winter (compared to 30 GW today) and 41 GW in summer (compared to 34 GW). No significant issue is thus expected on base load units. Peak load is expected to increase, following the evolution of power consumption, up to 48 GW in winter (compared to 38 GW today) and 78 GW in summer (compared to 63 GW). The ramp up of conventional generation would accelerate, as a response to the increase in combined renewable capacity. It ranges today below 7 GW (in summer) and is expected to nearly triple to 20 GW in summer. Balancing services are not expected to increase significantly, and remain in the range of 0.7% (summer) to 1.7% (winter). Similarly, ancillary services would not significantly increase, the ratio of synchronous machines remaining well above 40%. In summary, and provided the growth of the various renewable energies correspond to this forecast, Texas is not expected to see a drastic change of paradigm, but rather an increase in the share of its renewable energies (particularly solar). Mostly, the growth of renewable generation is expected to compensate for the growth of power demand (Fig. 4.39).

4.2.6.6   Australia—New South Wales

Transgrid experiences a clear peak load, which is more pronounced in summer, for this season is very hot. The ramp up is continuous over the day, and peaks at the end of the day. The profile is similar in spring, although less significant. The peak can reach around 15 GW in summer and no more than 12 GW in spring. The load can go as low as 6 GW at night (Fig. 4.40).

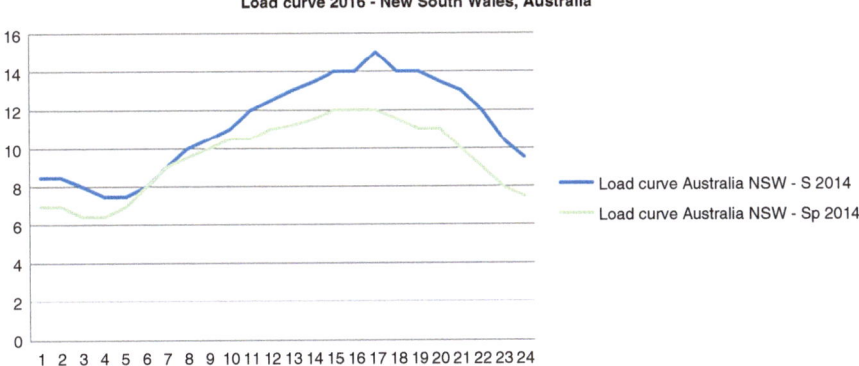

Fig. 4.40 Load curve 2016—New South Wales (Transgrid 2014)

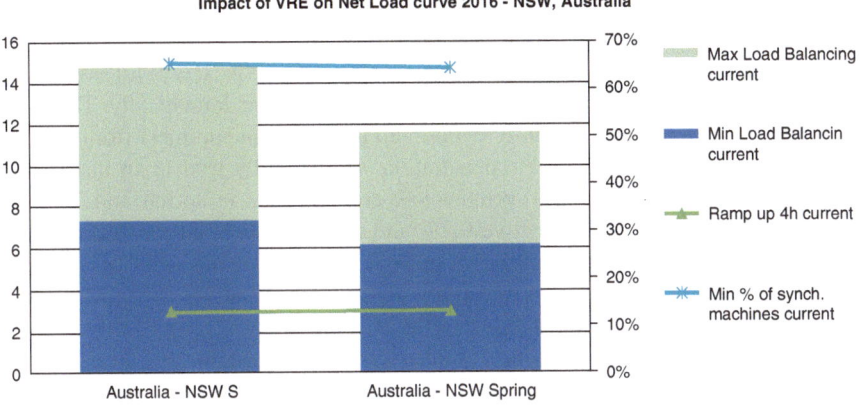

Fig. 4.41 Net load curve 2016—New South Wales (Source: Author's own calculation)

The load in New South Wales is projected to remain flat in 2030. Transgrid (2016) has indeed estimated in its annual transmission report a flat long-term trend until 2025.

New South Wales has today less than 2 GW of renewable power capacity, around 10% of its total power capacity (Transgrid/Renewable 2016). Variable renewable capacities are evenly distributed between wind and solar. The impact on the actual shape of the load is thus limited. There is a slight drop of the net load curve around midday because of solar, but it remains very low compared to the total demand. The effect is similar across seasons. Minimum load balancing ranges around 7 GW, and mid-peak and peak units account for another 7 GW or so. There is less than 3 GW of power needed to ramp up in 4 h, and the rate of synchronous machines exceeds 70% (Fig. 4.41).

The forecast of renewable development points to 4 GW of wind and up to 6 GW of solar by 2030, a significant share of the total power consumption.

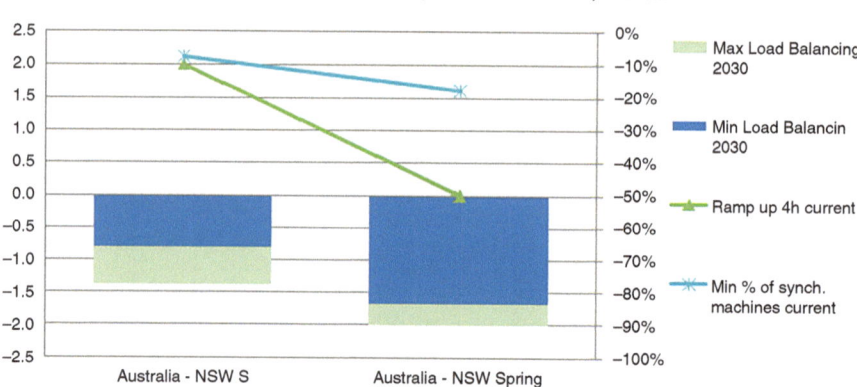

Fig. 4.42 Net load curve 2030—New South Wales (Source: Author's own calculation)

The minimum load balancing is expected to reduce slightly in spring. Since the maximum load required from conventional generation is expected to follow similar patterns, the amount of conventional units should not change dramatically. The ramp up is also expected to only slightly evolve, up to 5 GW (in summer) from today's 3 GW. The increase in renewable penetration will naturally lead to an increase in forecasting errors, which would remain low around 0.9% in spring and 2.1% in summer, compared to the respective 0.2% and 0.4% today. With the current forecasts of renewable penetration, and despite an increase in the need of ancillary services, the rate of synchronous machines on the network is not expected to drop below 40%, which means the stability of the network is not considered at risk.

With the current set of forecasts, it is thus not expected that the New South Wales network would undergo significant changes. Australia is a very good place in terms of solar irradiation, and therefore in terms of photovoltaic solar penetration. The further development of solar power, notably rooftop solar, could lead to a different set of assumptions and therefore important evolutions. Nevertheless, the shape of the load curve, in particular in summer, suggests that the volume of ramp up will be high in all scenarios, and that solar penetration will be significant. The New South Wales network is therefore expected to operate with a rather low volume of conventional base load units (Fig. 4.42).

4.3 How New Grid Challenges Disrupt the Traditional Economic Model

Traditionally, electricity markets have been made up of conventional generation which operated at various cost levels. Market deregulation in several countries in the past decades opened these markets to competition. Generation sources with low operating costs, such as coal and nuclear, generally kept operating as base load while other sources of supply, being more flexible by nature, operated in a non-continuous manner as mid-peak or peak load resources in order to meet the variation of demand over the day. The cost of electricity was calculated based on the cost of producing the necessary output at any time of the day. The power was distributed through the grid, a network designed to supply demand in the most optimized manner. Costs of maintenance and upgrades of the grid were added to the cost of electricity retailed. Transmission operators and distribution operators were kept as regulated entities as they were a key element to balance the network and ensure the security of supply.

The growing competitiveness of variable renewable energies such as photovoltaic solar and wind as well as the general regulatory support in the past decades have led to the fast increase in their share of the overall power generation mix. These sources of energy are by nature expected to operate to the maximum of their possible output. Their profitability is indeed conditioned by the initial cost of investment and the capacity factor that they are capable of achieving in operation. They can thus be considered as base load units, even though their production output is variable. For years, regulatory authorities and governments have encouraged their development by guaranteeing to investors an acceptable return on their investments, and forcing their integration into the supply mix. Now that their competitiveness is increasing, renewable energies do not need any more support from authorities in most places to compete in the market. Their fast development thus becomes natural, further increasing their competitiveness over time. This has led to a progressive evolution of the power generation mix. Traditional base load units, if found to be no more competitive compared to renewable energies, are decommissioned. This is the case of coal units when carbon prices are being applied, and this is as well strongly limiting the development of new nuclear power units. Many mid-peak units (some coal technologies, natural gas, etc.) also see their capacity factor being reduced by the increase in this base load, especially at times of the day where the renewable production is maxed out. The decrease of their capacity factor leads to profitability issues for these units. The generation market is thus structurally evolving as the share of variable renewable energies is increasing. Conventional base load units are being substituted by intermittent renewable energies. Already, the minimum load balancing (base load for conventional generation) could be zeroed in the coming decades in countries such as Germany. At the same time, the maximum load to be supplied by conventional units (in the absence of alternative technology) as a result of minimum production of renewable energies generally follows the demand load curve and its evolution upwards. This results in an economic equation that is impossible to solve with conventional generation remaining on the grid at much lower rates of operation. This will lead to increased volatility in wholesale prices, depending on the necessary power generation mix at a given time of the day. Prices would drop with a high share of renewable and could even reach zero at certain times of the day (because of the zero marginal cost of renewable energies); they would significantly increase at times when renewable output would be low. This will become a critical equation for the market organization since costs need to remain limited to a certain extent. Indeed, the main promise of renewable deployment is the promise of cheap, even "free" energy, with payment only for capacity. The whole-sale market will thus require in the short term to be strongly adapted to keep costs at a reasonable level. Retail markets could as well integrate more and more flexible ("time of use" or "real-time") pricing systems, in order to better reflect the variations in the cost of production during the day, as opposed to fixed retail prices in most countries of the world today. In addition, the growing load defection from the deployment of distributed generation (in homes and buildings) will reinforce the pressure on revenues for traditional actors interacting in the wholesale market.

The high penetration of renewable energies is also making the grid more complex to operate. By nature, these energies are difficult to predict. As traditional electricity markets are generally organized "one day ahead", it means that blocks of power generation are distributed among the suppliers 1 day in advance. Renewable pro-duction output is unknown at that time, and the errors in forecasts can reach up to 10–15%. In a low penetration scenario, this uncertainty is easily managed by the various operational reserves available. In a high penetration scenario however, this uncertainty leads to an important increase in the necessary balancing services, which are assumed by the operator. More reserve needs to be planned for in order to cope with the uncertainty of renewable production. While the volume of balancing services traditionally range below 2% of the average load supplied over the day, it could increase by up to 7% in some regions, leading to a large increase in system operation costs. This uncertainty could partially be mitigated thanks to better market accuracy. The creation of an "intra-day" market ("real-time" transactions) could help mitigate these issues. This would also represent an opportunity for operators to better distribute their capacity over a growing range of services markets, hence optimizing their revenues in the difficult context described above.

The stability of the network is also strongly connected to the number of synchro-nous machines connected to the grid. All conventional generators are synchronous machines and thus contribute to the stability of the grid, while photovoltaic solar and wind are not. With a high share of renewable energies in the production mix, the stability of the network could be undermined in some regions. Most regions however have not yet experienced such issues. If Spain and Germany are today ranging below 40% of synchronous machines connected to the grid at certain times of the day, their network interconnections to other European grids limits the problem. In remote regions with fewer interconnections, the problem could become paramount in the decades to come. The need to provide ancillary services to the grid in order to maintain stability is also included in the operator's duties. As a consequence, like for balancing services, they are factored into grid operation costs.

Finally, utilities are traditionally underutilizing their assets, for the reason that network assets are designed to support extreme conditions of operations, in order to guarantee supply at any time of the day, any day of the year. Since most of the time these conditions are not met, assets remain underutilized a large part of the time. Beyond ensuring security of supply, the role of transmission and distribution operators is therefore also to optimize the use of their assets, as well as to optimize the necessary investments in upgrades and extensions. These network assets have traditionally been designed several decades ago, in order to distribute energy in a "one-way" direction, from conventional generation to consumers. Conventional generation was traditionally located close to large consumption centers such as large cities, so as to limit the losses associated with the transportation of electricity, and directly connected to the transmission network. Now, renewable energies are traditionally distributed throughout the distribution network, a grid not designed to accommodate power generation initially. They are also located in places where weather conditions are favorable, and this often is not necessarily close to urban centers or large zones of power consumption, leading to a higher volume of losses, which costs are also borne by the operator. The profitability of network upgrades to better bear the additional costs associated with the penetration of renewable energies, in particular in the distribution network, presents its own issues in an already difficult context of asset underutilization. This could lead to a certain extent to the redesign of the grid infrastructure, from supplying energy to millions of consumers to connecting "islands" of self-generation and consumption (microgrids) together. In the short term, however, more investments are required, with lower profitability. Pressure on the profitability of grid operators' investments is thus rising with the penetration of renewable energies. A higher level of granularity and accuracy in distributing costs would help mitigate the inefficiency of the current market setup. Location based pricing could help, affecting the cost of transmission and distribution of energy to the actual generation asset operator, hence achieving better market accuracy.

The high penetration of renewable energies is thus leading in the short term to growing uncertainties and pressure on conventional players and the profitability of their assets, whether they be conventional generation actors or system operators. The market valuation of many utilities across the globe has as a result strongly dropped in recent years, showing the growing lack of confidence in the view that these companies are safe and profitable investments. The overall inefficiency of renewable penetration into the current grid thus has several consequences. First, the power generation mix has to progressively restructure around less base load polluting sources of energy, towards more flexible mid-peak load and peak load units. Natural gas (open cycle) and hydropower (wherever the potential has not yet been fully realized) are interesting technologies as they are extremely flexible to operate and not too sensitive to carbon pricing regulations, thus remaining competitive. Overall, wholesale prices will become more volatile to better reflect this evolution of the power generation mix. A greater level of flexibility will also be required in the market, through the growing importance of balancing and ancillary markets, and the introduction for instance of an "intra-day" market, inside one period of 24 h. Furthermore, grid operation costs if averaged today will require to be redesigned to give greater insight into the cost of integrating energy supply into the grid, notably through location based pricing. Renewable energies producing from remote locations (offshore wind for instance) have today a cost of electricity which only reflects the cost of production, while their cost of distribution is likely much higher than the cost of distribution for a conventional generator close to a large urban center.

Over time, electricity markets are thus expected to become more volatile, more real-time, and more geographically accurate. In this new regime, the LCOE will not be a perfect indicator of the actual cost of integrating a source of energy into the network. Nor will it be a good tool for understanding and monitoring the transition in progress. The changes in electricity markets will limit to a certain extent the main inefficiencies associated with renewable penetration: evolution of the mix towards more flexible but more expensive units, increased ramp up and peak capabilities, and a greater need for grid reinforcement. Regulatory authorities throughout the world today are reflecting on how to integrate these extra costs of operation into the competitive equation of electricity markets, meeting grid stability requirements while maintaining an acceptable level of prices for the final consumer. However, these measures might not be sufficient. The first promise of renewable energies, as already explained, lies in its zero marginal cost of production. If all the required power consumption could be met with renewable energies, the cost of electrical energy could theoretically approach zero. Consequently, as the penetration rate increases, conventional sources would progressively transform into a reserve and balancing infrastructure to cope with incidents. The "end game" model of electricity markets could thus reshape towards "free" energy supply, with a fixed charge for the infrastructure necessary to ensure the safe and reliable distribution of energy, a model very similar to the evolution witnessed in telecommunication markets. From a cost per unit of energy (or information) transmitted, the market would evolve towards a cost for secured access to a certain "bandwidth" of power (or information). Utilities would receive a fixed payment for developing and maintaining the overall "capacity" to meet the needs of the load demand. The recent development of "capacity" markets in various countries correspond to this trend. The "capacity" mechanism pays suppliers for making their capacity available at all times. Revenue from "capacity" has grown in all regions of the world where it has been deployed in recent years, and the development of "capacity" markets is currently spreading across countries. There is however today a debate in the community around its efficiency as a market tool. The second promise of renewable energies also lies in the potential of self and distributed generation. With distributed generation, the massive needs for transmission and distribution infrastructures are reduced. Instead, microgrids could be deployed, connecting to one another through the distribution and the transmission networks. These microgrids could be of different sizes, from one home to one city or region. Nevertheless, the principle would remain the same, and the sizing of the grid infrastructure would be revisited, from supplying the full amount of energy consumption required to providing the necessary bond between two "islands" of self-generation and consumption. The consequence of such development could be a significant redesign of the grid infrastructure and potentially a reduction of its costs.

Despite these long-term trends, the high penetration of non-flexible variable renewable energies calls in the short term for a greater flexibility from the rest of the market. If the restructuring of electricity markets is compulsory, regulatory authorities have to work on new market schemes in order to better value the flexibility required of the remaining actors. The new markets will be more volatile, more real-time, and more granular. Parallel to these evolutions, the development of "capacity" markets and microgrids represents the dawn of a possible complete market redesign, one where electricity is not valued any more in terms of energy, but in terms of power capacity—a complete change of paradigm for the industry. These changes obviously strongly challenge the role of the different actors of the market and their strategy going forward is scrutinized by market investors.

4.4 Main Regulation Challenges

The electricity markets present significant differences across regions. Historically, most were vertically integrated (power generation, transmission, distribution and retail) and regulated state-owned companies. In many regions of the world, electricity markets were partially deregulated in the last 20 years. The market unbundling has led to a number of private actors entering the market with additional financing capabilities. It has however been a complex exercise for authorities in deregulated countries regions to actually regulate operations in order to maintain proper balance and a healthy competitive environment. The International Energy Agency (2016) has mapped the specificities of electricity markets around the world (Fig. 4.43). Traditionally, all markets were vertically integrated regulated monopolies. They evolved first towards allowing independent power producers (IPPs) to coexist next to the vertically integrated utility. Then, unbundling split the utility into distinct companies (generation, transmission, distribution and/or retail). The wholesale market represented a further step with the creation of a specific generation market. Finally, retail competition has consisted of enabling competition for electricity retail at the final consumer point, the ultimate stage of competitive power markets. Most mature markets are already structured around a wholesale market and a competitive

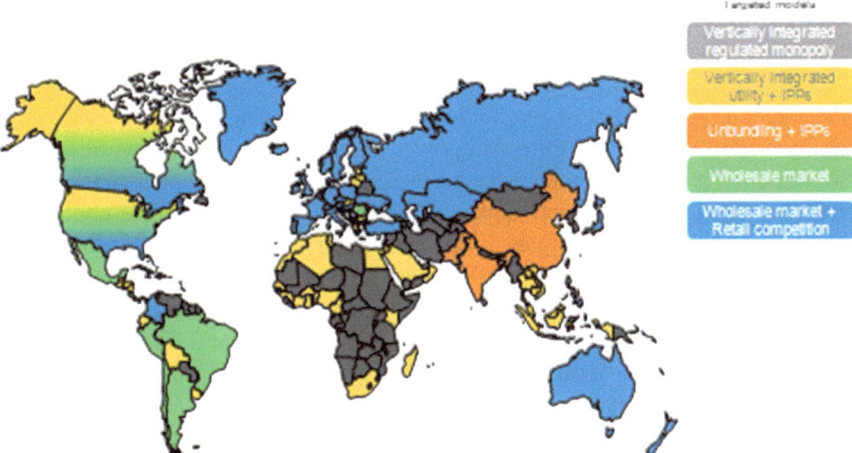

Fig. 4.43 Diversity of power markets (© OECD/IEA, Markets 2016)

retail market. China and India have already unbundled their utilities and allowed IPPs to operate. Most other countries are still heavily regulated, more or less tolerating IPPs. The situation in North America remains very specific, with a multitude of cases, depending on the state. Some states are already fully deregulated (e.g., California, Texas, Ontario); other have a wholesale market (e.g., Minnesota, Wisconsin, Tennessee, Alberta); yet other remain heavily regulated, tolerating IPPs (e.g., Nevada, Utah, Wyoming, Colorado, British Columbia).

With electricity markets significantly evolving, regulatory authorities around the world are striving to put together a coherent set of reforms which will help shape these markets in an appropriate manner, in order to attract investments while maintaining competitive electricity prices. They aim at encouraging the deployment of renewable energies in the grid and organize local market conditions in order to attract financing while ensuring improved reliability of operations. The progress of regulation however varies considerably from one region to another, both in terms of advancement and in terms of conditions. Part of the complexity of comprehending electricity markets' transition is that it is highly dependent on regulatory actions, and those vary significantly from one place to another.

In the United States, the Federal Energy Regulatory Commission (FERC) organizes the way electricity markets operate. FERC regulations are followed by local regulations at state level. It started with the Energy Independence and Security Act in 2007, followed by the American Recovery and Reinvestment Act in 2009, which led to an investment of 12 billion dollars to modernize the grid. Although most states have adopted legislation on distributed generation, fewer states have actually adopted legislations on advanced metering and demand response. Texas, California and the northeast states are ahead in the move towards smart grid development (EIA/Policy 2011). In 2016 the FERC started work on a new regulation (FERC745) which goes further in terms of setting up a demand response market. The Department of Energy (DOE) is also working on a number of regulations to be passed following FERC745.

In Europe, the Third Energy Package entered into force in 2009. The main objective was to ensure proper unbundling of the traditionally verticalized utilities in order to establish an open wholesale market as well as an open retail market, and encourage the deployment of renewable energies. Since then, the number of players has drastically increased in European power markets. Not surprisingly, difficulties have arisen. The new energy market package in preparation is intended to take stock of the experience from the previous years. It aims at facilitating the continuous penetration of renewable energies through price signals, and better integrating new players into the market, in particular flexible demand and energy service providers (European Commission 2016). One of the main challenges that the European Commission aims to solve is the creation of flexible markets by "offering consumers the possibility to actively participate in the market by adjusting their consumption to real-time prices, [by] ensuring that markets provide the right signals for investments in generation and the efficient use of available resources, [by] building missing electricity infrastructure and making better use of existing infrastructure, [by] ensuring flexible trading (. . .),[by] eliminating regulated prices on one side and inefficient support schemes on the other (...),[and by] introducing a more coordinated approach to renewable energy support schemes across Member States" (Europa 2015).

India is also gearing up smart grid deployment. By 2030, India should be the most populated country in the world, with above 1.5 billion people. It is also a country where access to electricity is among the lowest, with around 25% of the population without access to grid electricity. The Forum of Regulators in 2015 released a model of regulation, which remains to be adopted and possibly amended by local state utility regulators. The market for smart meters is expected to expand to 100 million units by 2020/2025. The Indian government has also committed to having 40% of energy (corresponding to 500 GW) come from renewable sources by 2030. The government has already put together a plan to deploy 175 GW of renewable energy within the coming five to 7 years, including 40 GW of rooftop solar energy (Engerati 2016). The vision of Indian authorities is to move from a weak centralized grid model to a more distributed model, with a high share of renewable energies, as well as microgrids which could run independently of the grid. Regulation in the various states of India is progressively evolving towards implementing this design for electrical networks.

China has also taken stock of the various initiatives around smart grid deployments. Being the topmost country in the world in terms of greenhouse gas emissions, it is also committed to significantly reducing its greenhouse gas emissions by around half of its 2005 level in terms of greenhouse gas level by unit of GDP. The Chinese government has defined five objectives around smart grid deployment: "increasing the utilization of renewable energy, improving grid asset utilization and management, increasing efficiency of operations and return on investment with reduced losses, upgrading and renovating technology to support comprehensive and intelligent power system development (...), and enabling demand-side interaction in optimizing the operation of the electricity system"(ETransition 2012). Now, the design of smart grid in China is essentially based on a "verticalized innovation" approach, which means it is run by state-owned utilities as part of 5-year plans. The State Grid Chinese Corporation (SGCC), the largest utility in China, plays a major role in turning the Chinese power system into a smart grid (IJEEP 2014). Consequently, there is little work on flexible markets and demand management programs, although time-of-use tariffs are already in place. Most programs aim to encourage the development of renewable production and reinforcing the grid, notably through high voltage and ultra-high voltage connections, and digitizing the grid through automation and software control systems at every level of the grid.

To sum up, regulators in every country are hard at work putting together a framework of rules for the enablement of smart grid technologies and for a faster penetration of renewable energies into the grid. Each country however differs in terms of its current situation and its strategic objectives, hence the pathways for implementing the transition of power systems will vary. As of today, more than 145 countries have set up renewable energy policies. Of these countries, 73 have feed-in tariff policies. A feed-in tariff is a payment made to households or businesses that generate their own electricity, and is the most widely deployed tool supporting the integration of renewable energy into the grid. Feed-in tariff policies are in place in 80% of high-income countries; in lower-income countries, the percentage is 65% (REN21 2015).

4.5 Summary

The load curve varies significantly across regions. First, some regions experience a significant peak load (e.g., Texas, New South Wales) while others have flatter curves (e.g., France, Germany, Spain) even though all experience a peak at the end of the day and, sometimes, early morning. This has a direct consequence on the power generation mix: base load is lower and there is a greater need for mid-peak and peak load units, which production costs are usually higher. There is also in regions with a high peak a natural incentive for introducing variable pricing, in order to incentivize people to limit their power consumption at peak times and to better reflect the actual cost of production of peak units. The load may also vary significantly across seasons. Electric heating leads to a strong increase in power consumption in winter in certain regions, while air conditioning has an important impact in summer in hot regions. For instance, France and California experience a significant difference in load between summer and winter, while Spain, New South Wales and to a certain extent Germany and Texas have a more constant load across seasons. These differences also yield different power market structures. When the load is different from one season to another, only a portion of the base load units can operate continuously across the year. Other units operate either during a part of the year, or the corresponding load is supplied by more expensive mid-peak units. The shape of the load curve across the year therefore determines to a great extent the mix of power generation in a given country. It also defines to which extent transmission and distribution assets are utilized, and explains why in many countries assets are significantly underutilized. It has been assumed above that the shape of the load curve will not change over years. However, usages evolve and so will the shape of the load curve. The progressive introduction of air conditioning leads to an increase in power consumption in summer in most areas and, in particular, a significant increase in the peak load. Electric heating when deployed also lifts up dramatically power consumption requirements in winter, in particular in places where winters are cold. The migration of heating from natural gas, coal or any other fossil fuel to electricity therefore leads to significant additional power requirements. The switching from electric heating to renewable heating (solar based for instance) could as well limit the increase in power consumption. The evolution and challenges associated with the transition are thus primarily based on the evolution of demand.

Then, the introduction of variable renewable energies also leads to major issues for the power market. Wind typically operates when wind is blowing. It thus can operate at any time of the day. Statistically, it operates more in winter than in summer. Solar operates only when the sun is shining, which means around midday, and drops down to zero at night. Solar is therefore more disruptive as it creates a gap for conventional generation demand around midday, while the same conventional generation is still required to meet the peak load in the evening. While wind power essentially modifies the minimum load balancing (with our set of assumptions), solar power also creates a ramp up phenomenon at the end of the day, when its production declines. This creates an additional need for mid-peak and peak units to be connected to the network. These units are more flexible, their production costs are more dependent on variable costs, and they can be operated only part of the day. Their cost of production is however usually much higher than base load units, leading to an increase in the volatility of wholesale prices. The minimum load balancing drop is expected to be significant in the United States (California), in Germany and, to a lesser extent, in Spain and France. The ramp up effect related to solar is expected to be particularly strong in California, Texas and Germany. Then, the massive introduction of variable renewable energy leads also to forecasting issues. Traditionally, forecasting errors on the demand do not exceed 2% of the average load. With important quantities of renewable energy on the network, supply forecasting errors can reach beyond 7% of the average load. This presents issues when forecasting demand, leading to a strong increase in balancing services required from the grid operator. This adds further cost for consumers. Another issue is the corresponding need for ancillary services required for maintaining network stability. It is generally admitted that stability issues remain negligible above a rate of 40% of "synchronous" machines connected to the network. Most power systems operate at higher ratios, especially when interconnected with larger networks. Now, the massive deployment of renewable energies can lead to a growing number of stability issues in weaker grids. Ancillary services could end up costing up to 10% of the total cost of production of electricity, according to the forecasts. Finally, another issue with the deployment of renewable energy is its impact on the utilization of grid assets. Traditionally, the network has been designed to supply important quantities of energy to large consumption hubs, such as large metropolitan areas. The supply was organized to minimize the transport of energy from production to consumption. However, renewable energies are located in places where they are expected to operate the most efficiently (seashores for wind, hot and sunny places for solar), irrespectively of the consumption location. This leads to an increase in grid costs. Additional investments may be required to connect those important sources of energy and convey it across the grid to the consumption centers. These additional investments may not be sensible from an economic standpoint. Reversely, self-generation within large urban areas can lead to significant underutilization of assets, putting their profitability at risk as well.

To sum up, the massive introduction of renewable power generally leads to a structural revenue issue for traditional operators: how to get reliable revenue for amortizing fixed costs and recovering the initial cost of investment. The resulting market inefficiencies also lead to increased costs of services for grid operators. On top of these issues, load defection from the deployment of distributed generation represents a further threat to the sustained existence of traditional operators. Market regulators have taken stock of these issues and developed new rules of engagement.

Overall, a greater flexibility of the system is required of all traditional operators to cope with the lack of flexibility and of predictability of renewable energies. Services markets are expected to develop, presenting interesting opportunities for flexible sources of supply to complement and optimize their revenues. Prices also need to become more granular. The development of "intra-day" markets ("real-time" transactions) helps mitigate the lack of predictability of renewable energies, while "location based" pricing helps better reflect the cost of transmission and distribution of different sources of energy. Going further, additional flexibility could be found at the consumer point, notably through the introduction of variable retail prices. In remote areas, particularly in new economies, microgrids represent an opportunity to mitigate the cost of grid reinforcement and build flexibility in a decentralized manner. As a result, if a certain level of restructuring is expected in a number of regions, the deployment of renewable energies will considerably increase the complexity of current markets.

The primary aspect of renewable energy in general is its zero marginal cost. Provided infrastructures are amortized, energy could become free. The identified market inefficiencies thus need to be considered as transitional issues towards an "end game", one in which electricity could become free from an energy supply (TWh) standpoint. The necessary system infrastructure, including conventional generation and the grid, would then be seen as a balancing infrastructure for providing reliable power supply at all times. While traditional markets are built on energy (TWh) transactions, the "end game" would correspond to markets built on capacity (GW). The balancing infrastructure overall would act as an insurance that power can be available at all times. This is similar to what happened in the telecom industry at the beginning of the twenty-first century, when the cost of communication evolved from a function of the volume of communication towards a fixed charge for accessing the network. The promise of renewable energies is therefore unlimited free electricity access against the payment of a fixed charge for connecting to the system. In a number of countries, market regulators have already started to develop capacity markets, which pretty much reflect this trend, even though their structure is still a hot debate topic. In such capacity markets, revenues from capacity has gone up constantly. Going forward, the distributed aspect of renewable energies could lead to the complete redesign of the grid infrastructure towards connecting "islands" of self-generation and consumption (microgrids) together. This evolution could potentially reduce the cost of the capacity infrastructure and increase the productivity of renewable deployment.

The role of tomorrow's utilities, in the new grid system which is shaping up, will thus be to maintain a balancing infrastructure optimized in terms of costs, and extremely flexible. Building flexibility into the system, from a time, geographical, and volume perspective, will represent the main challenge of all the actors in the market in the coming decades. The capability of the markets to reform and build flexibility will be essential to keeping prices low and preventing a surge of inefficiency costs. Solutions to achieve these goals exist; they will be reviewed in the last chapter of this book.

References

© OECD/IEA, Capacity Credit. (2011). *Modeling the capacity credit*. http://www. worldenergyoutlook.org/media/weowebsite/energymodel/Methodology_CapacityCredit.pdf

© OECD/IEA, Costs. (2015). *Projected costs of electricity*. https://www.iea.org/bookshop/711-Projected_Costs_of_Generating_Electricity

© OECD/IEA, Markets. (2016). *Repowering markets: Market design and regulation during the transition to low-carbon power systems*. https://www.iea.org/publications/freepublications/pub lication/REPOWERINGMARKETS.pdf

© OECD/IEA, Networks Infrastructures. (2013). http://www.iea.org/publications/insights/ insightpublications/ElectricityNetworks2013_FINAL.pdf

© OECD/IEA, Power Transition. (2012). *Securing the power transition*. https://www.iea.org/ publications/freepublications/publication/SecuringPowerTransition_Secondeedition_WEB.pdf

© OECD/IEA, WEO. (2012). http://www.worldenergyoutlook.org/publications/weo-2012/

© OECD/IEA, WEIO. (2014). https://www.iea.org/publications/freepublications/publication/ WEIO2014.pdf

© OECD/IEA, WEO. (2014). https://www.iea.org/publications/freepublications/publication/ WEO2014.pdf

50 Hertz. (2016). http://www.50hertz.com/en/Grid-Data/Vertical-grid-load

ADEME. (2016). *Contribution de l'ADEME à l'élaboration de visions énergétiques: 2030–2050*. http://www.ademe.fr/sites/default/files/assets/documents/85536_scenarios_2030_2050_syn theses_0613.pdf. See also at http://encyclopedie-dd.org/encyclopedie/territoires/scenarios-energetiques-ademe-2030.html

Amprion. (2016). http://www.amprion.net/en/demand-in-control-area

California ISO. (2013). https://www.caiso.com/Documents/FlexibleResourcesHelpRenewables_ FastFacts.pdf

California ISO. (2016). https://www.caiso.com/Documents/FlexibleResourcesHelpRenewables_ FastFacts.pdf

California ISO/Load Curve. (2016). http://www.caiso.com/outlook/SystemStatus.html

California ISO/Renewable. (2016). http://energyalmanac.ca.gov/electricity/electric_generation_ capacity.html

Clean Energy Wire. (2016). https://www.cleanenergywire.org/factsheets/germanys-energy-con sumption-and-power-mix-charts

De Felice, M., Petitta, M., & Ruti, P. (2014). *Short-term predictability of photovoltaic production over Italy*. http://arxiv.org/abs/1409.8202

DIW. (2011). *Balancing and intraday market design: Options for wind integration*. https://www. diw.de/documents/publikationen/73/diw_01.c.387225.de/dp1162.pdf

EIA/Policy. (2011). *Smart grid legislative and regulatory policies and case studies*. https://www. eia.gov/analysis/studies/electricity/

Enbala. (2013). *Grid inertia: An unseen hazard with renewable generation*. https://enbalatalks. enbala.com/2013/05/07/grid-inertia-an-unseen-hazard-with-renewable-generation/

Energy Almanac. (2014). http://energyalmanac.ca.gov/electricity/electric_generation_capacity. html

Energy Research & Social Science. (2014). *Electricity load profiles in Europe: The importance of household segmentation*. http://energyexemplar.com/wp-content/uploads/2014/09/Hayn-2014-Electricity-load-profiles.pdf

Energymag. (2016). http://energymag.net/daily-energy-demand-curve/

Engerati. (2016). *Smart grid progresses in India*. http://www.engerati.com/article/smart-grid-pro gresses-india

ERCOT. (2016). http://www.ercot.com/gridinfo/load/load_hist/

ERCOT/Forecast. (2016). *Long-term system assessment.* www.ercot.com/content/wcm/key...
lists/.../2016_LTSA_Update_6_21_2016.pptx

ERCOT/Projections. (2016). *Long-term load forecast.* http://www.ercot.com/gridinfo/load/forecast

ERCOT/Renewable. (2016). http://www.ercot.com/content/news/presentations/2016/ERCOT_
Quick_Facts_22216.pdf

Etransition. (2012). Wu Jiandong, Wu Jiang. *The smart grid in China: A discussion paper.* http://
www.etransition.org/images/2012.02.05.The%20Smart%20Grid%20In%20China.Final%
20Draft.pdf

Europa. (2015). *Energy: New market design to pave the way for a new deal for consumers.* http://
europa.eu/rapid/press-release_MEMO-15-5351_en.htm

European Commission. (2016). *Consultation on a new Energy Market Design.* https://ec.europa.eu/
energy/en/consultations/public-consultation-new-energy-market-design

EWEA. (2015). *Wind energy scenarios for 2030.* http://www.ewea.org/fileadmin/files/library/pub
lications/reports/EWEA-Wind-energy-scenarios-2030.pdf

Fraunhofer/Renewable. (2016). https://www.energy-charts.de/power_inst.htm

Greenpeace. (2015). *The Energy [R]evolution scenario.* http://www.greenpeace.org/international/
Global/international/publications/climate/2015/Energy-Revolution-2015-Full.pdf

Hirth, L. (2012). *Integration costs and the value of wind power.* http://www.eeg.tuwien.ac.at/eeg.
tuwien.ac.at_pages/events/iewt/iewt2013/uploads/fullpaper/P_181_Hirth_Lion_11-Dec-2012_
16:36.pdf

Hirth, L. (2015). *The optimal share of variable renewables.* https://www.mcc-berlin.net/fileadmin/
data/pdf/Publikationen/Lion-Hirth-2015-Optimal-Share-Variable-Renewables-Wind-Solar-
Power-Welfare.pdf

Holttinen, H., Meibom, P., Orths, A., O'Malley, M., Ummels, B. C., Tande, J., Estanquerio, A.,
Gomez, E., & Smith, J. C. (2008). *Impacts of large amounts of wind power on design and
operation of power systems* (NREL Conference Paper). http://www.nrel.gov/wind/pdfs/43540.pdf

International Journal of Energy Economics and Policy (IJEEP). (2014). Xiaoling Y., Jiangyang
Z. *An analysis of development mechanism of China's smart grid.* http://www.econjournals.com/
index.php/ijeep/article/download/753/433

InterRao. (2009). http://www.ebrd.com/downloads/sector/eecc/Baseline_Study_Russia.pdf

Intertek. (2012). *Household electricity survey: A study of domestic electrical product usage.* https://
www.gov.uk/government/uploads/system/uploads/attachment_data/file/208097/10043_
R66141HouseholdElectricitySurveyFinalReportissue4.pdf

IRENA/Forecast. (2014). *Renewable map.* http://irena.org/remap/IRENA_REmap_RE_targets_
table_2014.pdf

Jardini, J. A., Tahan, C., Gouvea, M. R., Ahn, S. E., & Figueiredo, F. M. (2000). *IEEE: Daily load
profiles for Residential, Commercial and Industrial low voltage consumers.* https://social.stoa.
usp.br/articles/0016/0239/REV2000009.pdf

MacGill, I., & Riesz, J. (2012). *Frequency control ancillary services: Is Australia a model market
for renewable integration?* http://ceem.unsw.edu.au/sites/default/files/documents/WIW13_
Riesz-FCAS-2013-09-02a.pdf

McKinsey/Forecast. (2016). *The role of natural gas in Australia's future energy mix.* http://www.
mckinsey.com/~/media/McKinsey%20Offices/Australia/PDFs/The%20role%20of%20natural
%20gas%20in%20Australias%20future%20energy%20mix.ashx

National Audit Office. (2014). *Electricity balancing services.* https://www.nao.org.uk/wp-content/
uploads/2014/05/Electricity-Balancing-Services.pdf

Nordpool. (2016). http://www.nordpoolspot.com/Market-data1/Power-system-data/Consumption1/
Consumption/Nordic/Hourly1/?view=table

NREL/Wind Integration. (2014). *Western wind and solar integration study phase 3.* http://www.
nrel.gov/docs/fy15osti/62906.pdf

NSW Government. (2015). *NSW renewable energy action plan annual report 2015.* http://www.
resourcesandenergy.nsw.gov.au/__data/assets/pdf_file/0008/586601/reap-annual-report.pdf

PLN. (2006). http://www.pln.co.id/dataweb/PLTA%20UPPER%20CISOKAN/EIA150311/EIA_Upper_Cisokan_Version_1.pdf

pv magazine. (2015a). *Germany will miss 2030 RE targets unless current policies change.* http://www.pv-magazine.com/news/details/beitrag/report%2D%2Dgermany-will-miss-2030-re-targets-unless-current-policies-change_100021916/#axzz4HIkRvL00

pv magazine. (2015b). *Turkey targets 5GW of PV by 2023 in new action plan.* http://www.pv-magazine.com/news/details/beitrag/turkey-targets-5-gw-of-pv-by-2023-in-new-action-plan_100018235/#axzz4HIkRvL00

PV Parity. (2013). *Grid integration cost of photo voltaic power generation, PV Parity.* www.pvparity.eu/results/cost-and-benefits-of-pv-grid-integration

Red Electrica. (2016). https://demanda.ree.es/demandaEng.html

Red Electrica/Renewable. (2013). http://www.ree.es/sites/default/files/downloadable/preliminary_report_2013.pdf

REN21. (2015). *Global status report.* http://www.ren21.net/wp-content/uploads/2015/07/REN12-GSR2015_Onlinebook_low1.pdf

Renewables International. (2014). *German electricity demand forecast for 2014–2019.* http://www.renewablesinternational.net/german-electricity-demand-forecast-for-2014-2019/150/407/83485/

RTE. (2016). http://clients.rte-france.com/lang/an/visiteurs/vie/courbes.jsp

RTE/Demand. (2015). *Bilan prévisionnel 2015.* http://www.rte-france.com/sites/default/files/bp2015_synthese.pdf

RTE/Renewable. (2015). http://www.rte-france.com/sites/default/files/panorama_des_energies_renouvelables_2015.pdf

RTE/Usages. (2014). http://clients.rte-france.com/htm/fr/mediatheque/telecharge/bilan_complet_2014.pdf

Statnett. (2016). http://www.nordicenergyregulators.org/wp-content/uploads/2013/02/Nordic_Market-report_2013.pdf

Statnett/System. (2015). *Statnett system operation and development plan 2014–2020.* http://www.statnett.no/Global/Dokumenter/Kraftsystemet/Systemtjenester/SMUP%20Overview.pdf

STORE Project. (2016). www.store-project.eu/documents/target-country.../en.../energy-storage-needs-in-spain

Taylor, J., de Menezes, L., & McSharry, P. (2006). A comparison of univariate methods for forecasting electricity demand up to a day ahead. *Science Direct.* http://eprints.maths.ox.ac.uk/281/1/ijf22p1y2006taylor.pdf

TEIAS. (2008). http://www.teias.gov.tr/eng/ApkProjection/CAPACITY%20PROJECTION%202009-2018.pdf

Tennet. (2016). https://www.tennettso.de/site/en/Transparency/publications/network-figures/annual-peak-load/maximum-voltage-network-load-curve

Tepco. (2001). http://scholarship.claremont.edu/cgi/viewcontent.cgi?article=1075&context=scripps_theses

Tielens, P., & Van Hertem, D. (2012). *Grid inertia and frequency control in power systems with high penetration of renewables.* https://lirias.kuleuven.be/bitstream/123456789/345286/1/Grid_Inertia_and_Frequency_Control_in_Power_Systems_with_High_Penetration_of_Renewables.pdf

Transgrid. (2014). https://www.transgrid.com.au/news-views/publications/network-development-strategy/Documents/Network%20Development%20Strategy.pdf

Transgrid/Renewable. (2016). http://www.resourcesandenergy.nsw.gov.au/__data/assets/pdf_file/0008/586601/reap-annual-report.pdf

Transgrid/Transmission Report. (2016). *Annual transmission report.* https://www.transgrid.com.au/news-views/publications/Documents/Transmission%20Annual%20Planning%20Report%202016.pdf

Transnet BW. (2016). https://www.transnetbw.com/en/key-figures/load-data/load-curve?app=last
    verlauf&activeTab=graph&auswahl=day&date=01%2F12%2F2016&selectMonat=0&select
    Jahr=2016
Wind Power Monthly. (2015). *Analysis: California wind power could increase by 21GW by 2030
    updated.* http://www.windpowermonthly.com/article/1366920/analysis-california-wind-power-
    increase-21gw-2030-updated
World Economic Forum. (2015). *The future of electricity.* http://www3.weforum.org/docs/
    WEFUSA_FutureOfElectricity_Report2015.pdf
WWF. (2014). *Turkey's renewable power.* http://awsassets.wwftr.panda.org/downloads/wwf_tur
    key___bnef___turkey_s_renewable_power___alternative_power_supply_scenarios_until_.pdf

Chapter 5
Solutions for a Sustainable Transition

Electricity markets are undergoing a spectacular transformation, which final outcome yet remains to be seen. The first transition involves demand. Worldwide demand is set to almost double by 2035, although many differences exist across regions. Most of the growth is expected to come from non-OECD countries. There, economic development is driving electricity consumption. In OECD countries, growth is expected to be slower, and in some cases even negative for some period of time due to energy efficiency measures. Usages of demand will also evolve. The further migration of heating to electric heating, the development of air conditioning, the fast deployment of information and communication technologies, and the development of electric transportation could all lead to further significant increases in consumption. In addition, it would radically modify the shape of the load curve, leading to more consumption, and more peaks for the system to handle.

The second transformation is associated with the rapid expansion of renewable energies, in particular variable energies such as wind and photovoltaic solar. The deployment of renewable energies leads in many regions to the restructuring of power generation, with fewer conventional base load units and more peak units. The development of competitive renewable energies also leads to the expansion of microgrids, an interesting alternative to grid power, both from the end consumer standpoint when prices are high, or for the utility, when it helps defer grid investments or sustain grid balance (JISEA 2015). Renewable energies also yield increased grid costs. The cost for balancing services increases because of the predictability issues associated with renewable power. Ancillary services also increase because of grid stability requirements. The penetration of variable renewable energies also leads to important grid reinforcement needs for coping with multi-directional flows of energy, which modify deeply the original architecture of the electrical grid. This also leads to a considerable increase in the possible failure modes of the grid, making grid management more complex. All these changes happen at a time where increased investments in the grid are becoming more difficult to realize. The risk profile of electric utilities has indeed significantly evolved in recent years because of the uncertainties associated with new electricity markets, and

© Springer Nature Switzerland AG 2019
V. Petit, *The New World of Utilities*, https://doi.org/10.1007/978-3-030-00187-2_5

also because it was once considered a secured, regulation-driven and stable market. All this is changing now with market unbundling. In addition, the politically-driven limitation of electricity price increases makes it more difficult for utilities to actually guarantee return on investment (in many countries, electricity remains subsidized). To summarize, electricity markets are under considerable transformation. It has long been stable, with a few actors playing in a regulated market and managing a well understood grid with slow evolutions. The market now sees the emergence of a multitude of actors, a deregulated market organization which has yet to mature in a number of countries, and a grid which has become more complex to operate with evolutions more difficult to forecast and predict. These changes obviously lead to the restructuring of the traditional market players, in particular fossil-fuel based generation. This restructuring is inevitable.

It has also become more difficult for grid operators to maintain a high level of reliability with the higher penetration of renewable energies on the network, and utilities suffer from increased system costs which reflect the inefficiency of traditional grid management. The challenge that grid operators face is thus to maintain over time a high level of reliability and efficiency in order to keep a sustainable performance, which will ensure customer loyalty, economic stability and competitiveness, as well as sustained affordable financing.

Renewable operators' profitability highly depends on initial investment costs as well as the capacity factor of their renewable power plants. The ability of these operators to maintain a high level of production despite operating conditions (especially considering the lack of predictability of renewable energies) and to maximize their revenue is thus critical. Additionally, the excess of renewable energy available on the network at some times of the day leads to a number of issues. First, the price in case of excess of supply tends to drop, reducing the benefit of operators. Then, in case of excess of supply, market regulators are today considering new mechanisms such as power curtailment. This could strongly impact renewable operators' bottom line if they cannot react to the new mechanisms. Market regulators are striving today to define the right scheme for all, with each country having its different recipes.

Finally, consumers have so far not been concerned to a great extent by the transition at play. More volatile prices yield a natural incentive to optimize their consumption in order to minimize the electricity bill. Retail price regulations are evolving to better reflect the important variations of the cost of production at different times of the day. Consumers are thus getting progressively more engaged with the transition. Additionally, the sharp drop of distributed energy costs, such as rooftop solar energy, leads them to engage in electricity markets as potential producers of energy on the grid, not simply as consumers. Finally, the demand side clearly represents today the greatest potential source of flexibility, a critical factor for running an efficient and affordable energy system. Demand side management, enabled by distributed generation and storage, could become the cornerstone of this new energy system paradigm.

In this context of increasing difficulties, digitization shows promise as a key enabler that can help producers and grid operators reach a new level of performance. A debate in the IT community has emerged recently on the distinction between digitization and digitalization. Digitization refers to the process of turning analog systems into digital form, while digitalization refers to the process of changing business models with the help of digital systems. We will refer to both as digitization. We indeed consider one is necessary to the other. Digitization brings both the capability to interact real-time with all components of the network, as well as the access to a multitude of data which up to now were unknown or accessible only once in a while. These new capabilities to monitor the complete grid (down to the lowest consumer point) digitally in real time help improve drastically grid management and the operation of the power market. Real-time information and control capabilities also engender faster and less encumbered reactions to incidents and outages. Efficiency is improved by optimizing operational margin reserves which are required to secure reliability of supply. The capability to better predict load and supply forecasts that digitization provides helps improve capacity factors of renewable plants, as well as the network configuration to minimize additional costs and optimize network planning to tame or defer grid investments. Finally, real-time digital interaction is made easier among market players, leading to a more fluid and flexible market, with increased interactions between consumers, producers and grid operators. McKinsey (2016) estimates that digitization applied throughout the value chain could yield up to 6.6% EBIT increase in generation efficiency (through optimized maintenance and fuel management), 2.5% in trading improvements (through better energy balance and decision making), 4.3% in distribution efficiency (less losses, optimized maintenance and workforce management), 8.5% in retail performance (with new products, prices and customer segmentation), and 1.3% overall in enterprise resource planning.

The digitization of the grid yields a "smart grid" that then becomes a key enabler of the transition towards a new energy landscape. In essence, digitization enables more flexible operations by the different actors of the market.

5.1 Digitizing the Distribution System

The digitization of the grid starts with grid operators. Transmission system operators (TSOs), who are responsible for maintaining network balance under all conditions, have long digitized the transmission network. Traditionally, power generation is connected to the transmission network. TSOs thus need to balance production and its distribution to distribution system operators (DSOs), which connect the millions of end-user consumer points. TSOs have thus to manage a complex interconnected network (including interconnections with other countries) with various generation sources and a multitude of consumption points. All primary substations and connection points have long been automated with digital protection and control systems. These substation automation systems are connected to a global control center, which manages the overall network stability and the communication across the various sources of supply. DSOs, though, have not fully undergone this transformation. For the most part, distribution systems have remained unautomated. The role of DSOs has mainly been to maintain the huge base of network assets. When incidents occur on the distribution network, DSOs are mostly informed through customer calls, and then dispatch teams on site to restore power at the fault location (U.S. DOE/ADMS 2015). Renewable energies are mostly connected to the distribution network. The complexity of grid operation modes at the distribution level has thus been tremendously rising over the last few years, and shall become increasingly more complex over time. DSOs now experience a variety of situations and operating modes which mirror what TSOs experience. The complexity they face is even greater because of the level of expansion of the distribution network, the number of generation sources connected to the network, as well as the millions of consumer points with different consumption profiles. So DSOs are facing a major disruption in the way they have to operate.

Traditionally, DSOs have some level of automation in their primary substations, where the transmission and the distribution networks connect together. However, secondary substations were only automated (to a lower level than primary substations) only a decade or so ago. Many distribution networks across the world remain unautomated. Secondary distribution is key as it converts medium voltage into low voltage for further distribution to consumer points. These substations, traditionally equipped with simple switches, now experience power flows in all directions, unlike before. Automation of those parts of the network is thus critical to better understand the reality of power flows. The deployment of medium voltage/low voltage (MV/LV network automation is now rapidly expanding throughout the world. This represents an important investment for electric utilities as there are considerably more secondary substations than there are primary substations. From secondary substations, the next lower level is the end-user consumer point. Traditionally, end-user consumer points are equipped with meters, which record the electricity consumption. These meters are checked on a regular basis by DSO operators. The advent of smart metering helps digitally record this data and avoid regular visits from the DSO operator to every home. Smart metering systems are able to record power consumption on a very regular basis, for instance every half an hour. The resultant huge amount of data helps DSOs draw accurate load profiles of each and every consumer point and run statistical analysis on consumer behaviors, which then can be turned into business development programs for the utility retailers so they can work toward a higher level of flexibility, increased energy efficiency, etc. It can also help better understand the network operating model, by providing real-time data; traditional grid models only estimate the actual load, since no data is available for monitoring. Finally, the data collected by smart metering systems can be integrated into the global power outage system architecture to monitor actual power restoration in case of incidents (Sempra Energy 2015). With the deployment of digital technologies, the volume of data that a DSO has to consolidate is now being multiplied by a spectacular factor of 10,000! Traditional DSOs thus face a major disruption to their way of working. The competencies they need to efficiently meet the challenges ahead are also changing dramatically.

From digitization of the distribution grid, the next step is the automation of local systems in a decentralized manner. Simple subareas of the network can be automated to provide faster response to incidents. Self-healing grids (Schneider Electric/SHG 2016) can reconfigure automatically meshed networks in case of incidents, providing power restoration in a matter of seconds, hence improving the overall network reliability without 24/7 constraints on operation teams. Local automation also improves the control of voltage at different nodes of the network through volt/var. applications (U.S. DOE/VVO 2009). Volt Var Optimization (VVO) indeed helps adjust voltage and reduce losses in feeders. With the emergence of distributed renewable energies, it has become more important to maintain voltage within acceptable boundaries for the load as well as minimize losses on the network. The potential for peak demand reduction could reach up to 2.5%, according to estimates from the U.S. Department of Energy (2009). Local automation also helps adjust the protection plan of the network in real time, taking into account the variations of the system characteristics introduced by variable renewable production levels. The final application of decentralized automation has to do with microgrid systems. A subarea of the grid can be operated as an autonomous network with its own power generation and consumption points. It is integrated into the wider network, with which it exchanges power flows. The development of decentralized energy resources (DER) such as wind and solar has led to the rapid expansion of microgrid systems. Electric utilities are often interested in developing microgrid systems, especially in areas where the cost of grid reinforcement to supply remote areas is not competitive. Operating subareas of the network as independent microgrids which can (or not) be connected to the larger grid can be an economical alternative. Many consumers are also interested in developing their own generation capabilities in order to limit the energy they consume from the grid. It is often the case in large industrial facilities which produce secondary products such as heat or gas, and reuse them in order to produce power. Microgrids allow them to minimize the power they consume (and buy) from the grid. The development of cheap renewable electricity has accelerated the development of such systems around the world. The United States, where many states experience volatile electricity pricing, have long been sensitive to optimizing their electricity consumption. The country is today the foremost market for microgrids (Navigant 2016), although the concept is developing fast elsewhere in the world, in particular in new economies where access to electricity remains an issue.

Data from the automation of local systems points can be transmitted to a centralized control system which represents the brain and heart of utility operations. A number of functions can then be realized using this massive amount of information available. There are essentially four groups of applications which a utility uses to drive its operations.

The first group corresponds to network management applications. They usually consist of a number of tools, including SCADA, Outage Management System (OMS), Advanced Distribution Management System (ADMS), and Distributed Energy Resources Management System (DERMS) (Schneider Electric/Utility Solutions 2016). The SCADA interacts in real time with all devices connected on the grid. It collects the data and can control the network actuators remotely. With its basic topology, the electric SCADA helps the operator to remotely control the grid.

The OMS helps manage incidents more rapidly and accurately. It evaluates the impact of an outage on customers, prioritizes responses, and assigns crews for fast restoration. It also helps communicate to consumers root causes of failure and response times expected for restoration. The ADMS is a set of analytics which help a utility understand and control the grid efficiently. It covers a variety of functions: network operations control, network operations optimization, network operations analysis, network development planning and training. Network operations control performs the load flow of the network and provides fault management, supply restoration, and load shedding. Network operations optimization includes voltage control and volt/var. optimization, network reconfiguration, short-term load forecasting and load management, as well as operations improvement algorithms. Network operations analysis provides a variety of analytics around energy losses, operational losses, as well as harmonics analysis, security assessments, fault calculation and relay protection analysis. Network development planning provides tools to better plan long-term load evolution, network development, and network reinforcement needs. Finally, training features help simulate and train operators for various situations as well as develop and refine emergency procedures and courses of action. The DERMS helps manage renewable sources as a portfolio of energy to better anticipate the impact of their supply on the network, anticipate market participations, and secure grid support. These systems also often use the contribution of weather forecasting systems to help plan for weather evolutions and their impact on possible outages (OMS), distributed energy generation (DERMS, ADMS) and load forecasting (ADMS).

The second group of applications covers management applications for utility assets. Asset management for utilities traditionally covers a very wide scope across the full enterprise. The British Standard PAS 55:2008 has historically set the foundation for the definition of an international ISO 55000 standard on enterprise asset management. A number of principles apply to this discipline. Asset management is first expected to be holistic and cross disciplinary, with the focus on the global value that the asset brings to the business; it is expected to be risk based, "incorporating risk in decision making"; it is expected to be sustainable (and optimal), delivering best asset lifecycle and system performance while optimizing the cost/performance equation (IAM 2016). There are a number of asset types. There can be intangible, financial, informational or human assets, leading to a very wide definition of asset management. There are three main levels of asset management. The first one looks after managing the asset lifecycle and optimizing it (IAM 2016). It controls the creation/acquisition, utilization, maintenance, renewal/disposal of the asset. The second level monitors and controls the asset systems' performance over time, optimizing risk levels and costs. Finally, the third level takes care of capital investment optimization and sustainable long-term planning of asset portfolios (IAM 2016). The scope of asset management thus embraces vertically many topics from asset lifecycle operations to planning, and horizontally from intangible assets to human and financial assets. With such a wide scope, it is not surprising that a variety of software applications have flourished. Today, only a handful of large software companies offer enterprise asset management systems (Oracle 2014; IBM 2016) but it is often difficult to deliver true knowledge of the actual asset lifecycle and behavioral particularities. Large IT companies often bring a tremendous efficiency gain through a restructuring of internal processes and operations. The delivery of extended value has remained the hurdle because of the difficulty of grasping fully the technical and behavioral particularities of electro-mechanical assets. Still, various specialist software specific to the electricity industry exists, and utilities strive to get the most out of each. We can define two main levels of asset management for utilities. The first level, called Operational Asset Management (OAM), provides specialized understanding of electro-mechanical asset lifecycle. They provide condition monitoring, asset health check, preventive maintenance recommendations based on the utilization of assets, as well as predictive maintenance features based on big data analytics. This first level then connects to the second level, Enterprise Asset Management (EAM), which provides all features as described in ISO 55000, such as maintenance schedules, workforce management, asset system optimization, capital investment optimization and sustainability planning. The first system thus serves as a "specialized" input to the second system. A third system often connects to these applications. The Geographical Information System (GIS) provides a unique base of asset data, geospatially referenced, which can be used by the OAM, as well as by the network management suite.

The third group of applications has to do with smart meter deployment. There are millions of smart meters deployed in residential applications. These meters are read every 15–30 min so there is a massive amount of data which needs to be collected. This information needs also to be properly handled. The time at which it is collected needs to be accurately known, and missing data (in case of communication interruption) needs to be estimated based on past data. The Advanced Metering Infrastructure (AMI) is the system in charge of collecting this data. It is often associated with a Meter Data Management (MDM) system which controls the accuracy of the data. Both provide a large base of meter data which can be exploited further by a variety of customer analytics. These analytics provide such features as billing, customer profiling and segmentation, and serve as a base for designing new retail services. The data from meters can also be used by the network operations management suite, in particular the ADMS, for better accuracy of the network model which is at the core of all the ADMS analytics around losses, reconfiguration, configuration optimization, and planning. Finally, a portion of the data from meters can be used for demand response programs. The data indeed gives insights into the consumption in each home, and therefore can be used for this purpose. When distributed generation is installed, it also provides information on both generation and consumption. Now, demand response programs are often required to look beyond the meter to turn on/off load points such as heating or cooling. In such cases, the actuators are "beyond the meter" and require a specific infrastructure to be put in place. This granularity is in general not available from smart meters. Often, the Demand Response Management System (DRMS) corresponds to a different application which uses only part of the data captured by the meter. Smart metering infrastructure is the responsibility of the DSO so data from smart meters can be used by the DSO as an input to the ADMS. However, this data is essentially used by retailers in their commercial relationship with their consumers. Retailers are the ones who derive from smart meters information on a vast array of commercial services, from billing to special offers and demand response programs (McKinsey/Digital Utility 2016). Smart metering deployment has essentially been pushed by regulatory authorities. They have been widely deployed in a number of countries, although large countries such as Germany have so far refused to move forward with their deployment, with the concern about the exact value added to the already complex digital ecosystem of utilities. Smart meter investments are indeed massive for utilities (several billion dollars) and therefore hamper other investments in the grid. So far, the entire smart metering infrastructure has been conditioned by the selection of the smart meters' vendor. It has not been possible for most utilities to actually choose among several smart meter actors and to diversify their procurement. Generally, one vendor is selected for the entire infrastructure because of the complexity of the communication infrastructure and the specificities of each meter manufacturer. There are very few players who have been able to propose a meter-agnostic metering infrastructure (Schneider Electric/AMI 2016).

The last group of applications corresponds to supporting tools which are essential to the overall operation of the utility IT landscape. The GIS provides a geospatial view of referenced assets and is thus a critical element for asset management applications, ADMS and OMS network management applications. The Work Order Management (WOM) application helps manage switching procedures and all activities related to planned maintenance. It also consolidates related work and safety documents, stores data, etc. It is therefore a key application to organize the work of field crews. It can be associated with the Mobile Workforce Management system (MWM), which schedules dispatch of field workforce and provides all information available for execution of the defined procedures. Finally, a variety of more general IT systems connect to this digital ecosystem, such as the Customer Information System (CIS), which stores the data from customers, the Customer Relationship Management system (CRM), which monitors the commercial and contractual relationship with customers and, more generally, the Enterprise Resource Planning (ERP) system of the utility.

The digital utility landscape is summarized in Fig. 5.1.

One of the main issues utilities face when going digital is the convergence towards a unified digital ecosystem where data is always accurate, permanently up to date, and entered only once in the whole ecosystem (one version of the truth). Such convergence is a paramount requirement considering the variety of tools using this data, the criticality of the applications running on those assumptions, as well as the volume of data being handled. Now, it turns out that often utilities own part of the overall landscape. Often these systems come from different vendors and have been deployed over time, with not necessarily a global enterprise software architecture in mind. As a consequence, the data is often inaccurate and the systems do not connect together well. As a result, management of the digital infrastructure is extremely cumbersome, and prone to errors. On top of that, the lack of connectivity across systems leads to a reduced value added from these applications. Most utilities thus face the critical challenge to converge from where they stand towards a unified IT

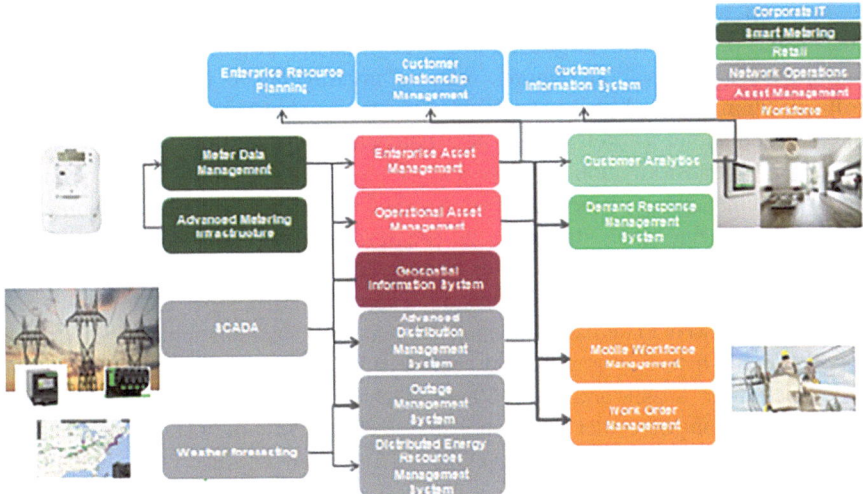

Fig. 5.1 The utility digital landscape

infrastructure which consolidates data in a secure and accurate manner and makes the most of each system while providing bridges across them to generate increased value. This data integration issue is a major challenge in most utility applications and requires a tailor-made approach with a specific roadmap which will differ from one utility to another. The progressive convergence of various elements of this system to standards helps, as does utilities not relying on a single vendor for all their applications. IEC 61850 has already become the standard in network automation products and systems. The CIM standard already widely applies across utility IT systems and needs to be more generalized. On their part, standardization bodies are working on framing the scope and interfaces of each application in order to bring some order to the overall landscape.

The deployment of such vision in a utility requires a massive investment. Cloud based technologies enable more and more the use of software as a service (SaaS). SaaS technologies apply well to utilities, provided the critical data handled is secured properly. SaaS technologies correspond most of the time to a less intrusive alternative to the existing landscape of utilities. It is also a cheaper solution, since the hardware required to handle the system does not need to be purchased using capital expenditure budgets. Finally, it requires less understanding of the specificities of IT systems, so there is less need to acquire expensive technical talent. Last but not least, for the many utilities which have already embarked on their digital journey, the choice of software as a service is an efficient way to pay for value, to buy time to change internal processes, and put management structures and operator competencies in place.

In summary, DSOs face a major disruption in the way they operate electrical networks. The level of complexity of their operations has increased considerably with the fast penetration of variable renewable energies. At the same time, they need to keep up with increasing requirements in terms of reliability, efficiency and sustainable performance. Digitization is a key enabler to reach this new level of performance. However, it requires massive investments in the automation and software domains, and also calls for a deep change in how DSOs manage their internal operations as well as the new competencies they now need to master. This digital transformation has proven to yield huge benefits when properly deployed. One among many examples, Austin Energy has been able to secure very ambitious strategic goals thanks to its progressive move to a digital ecosystem (U.S. DOE/ADMS 2015). The city of Austin, Texas has defined six main strategic goals: the deployment of 35% of renewable energy by 2020, 20% reduction in CO2 emissions, 800 MW of demand side management, while maintaining a price increase for electricity below 2%, reliability below 0.8 (Standard Average Interruption Frequency Index SAIFI) and below 60 min (Standard Average Interruption Duration Index SAIDI), and customer satisfaction above 80%. The current status on each one of those strategic goals (U.S. DOE/ADMS 2015) has proven Austin Energy to be on the right track to meet its ambitious 2020 goals.

5.2 Optimizing Revenues for Renewable Operators

Renewable energies are dispersed across the network in a multitude of small capacity power plants, a supply architecture very different from the traditional one, which is made of large conventional power plants connected to the transmission network. Variable renewable energies also lack predictability and face issues in maintaining fixed and scheduled generation since their input is by definition variable and dependent on weather conditions. As a consequence, they create a number of issues on the grid which were covered in the previous chapter. Notably, renewable energies lead to increases in generation ramp-ups on the grid, balancing and ancillary services, as well as grid reinforcement costs. Now, the specific context of the grid is important to understand the exact impact of variable renewable energies connected to the network. The size of the power system, the level of interconnection with other grids, the geographical distribution of renewable generation, the actual characteristics of conventional generation, the market size, the various operational practices and finally the expected evolution (energy planning) are all elements which strongly influence the impact of renewable energies on the grid (Irena/Grid codes 2016). In traditional utilities, power generation, transmission, distribution and retail activities are combined. The choices made in terms of generation or grid investments are based on the optimization of grid operations while ensuring the best possible security of supply. In countries where the markets are unbundled, generation, transmission, distribution and retail are run by different companies. There, overall system optimization is more difficult to achieve. In these countries, grid codes were established in the last decades in order to ensure an overall system operation which continues to meet the demanding standards of reliability at an optimized cost of supply and distribution. These grid codes include planning codes, operating codes, connecting codes, data communication codes and balancing codes (Eurosunmed 2015). Planning codes refer to upcoming investments. They mainly relate to generation and networks (Irena/Grid codes 2016). Operating codes correspond to various rules to ensure a stable supply of energy. These rules relate to operational security, operational planning and scheduling, load frequency control and reserve, as well as emergency procedures (Irena/Grid codes 2016). Connecting codes refer to interconnections between different components of the grid and different grids. Data communication codes correspond to rules to be respected in order to ensure a sustainable and seamless exchange of real-time data among the actors of the grid. Balancing codes refer to roles and responsibilities and market rules related to grid balancing. The third package of European regulations includes rules such as capacity allocation and congestion, forward capacity allocation, electricity balancing, requirements for generators, demand and HVDC connections, operational security, operational planning and scheduling, load frequency control and emergency and restoration procedures (Eurosunmed 2015).

Grid codes thus encompass a number of rules that renewable operators need to follow. These rules correspond to a number of demanding technical requirements for renewable operators to be authorized to supply energy on the network. Renewable plants are required to operate at a certain period of time when the network conditions are outside certain boundaries of voltage and frequency, in order to avoid further imbalance of the network. They are also required to provide frequency and voltage support under certain conditions, and to operate under certain conditions of power quality. Their behavior in case of fault (called Fault-Ride-Through (FRT)) is also regulated. They are not expected to disconnect in case of voltage dips for instance, and are required to provide voltage support by means of reactive power production. Some limitations exist as to the variations of power output they are expected to work with in order to limit ramp-up and ramp-down of conventional power capacity, which increases balancing services costs. Renewable energies can also be called upon to help mitigate network congestion, which requires power re-dispatch (Irena/Grid codes 2016). Despite some agencies doubting the complete usefulness of such requirements on renewable sources connected to the distribution network, they also acknowledge that grid codes are only a few decades old in most countries, and still require some time to adjust to (Eurosunmed 2015). As the share of renewable energies increases on the grid, the requirements published in the grid codes also tend to become more demanding.

In addition, market rules within which renewable energies operate can also vary across countries and lead to different constraints on renewable operators. In Spain for instance, renewable operators are made responsible for balancing costs, which is not the case in Germany (Fichtner 2010). The wholesale market is generally organized as a "day ahead" spot market. Some countries are moving to a more continuous spot market, something renewable operators have to cope with. The overall system benefits from this since balancing services are reduced. The lack of predictability of renewable energies is more easily managed from a market and trade point of view if transactions happen closer to dispatch. Most countries have organized various mechanisms to balance the grid and maintain reliability. These mechanisms present interesting opportunities for renewable operators, although they are primarily designed for conventional power generation. In a number of regions, location based pricing is emerging to reflect more accurately the cost of integration of energy. This takes stock of the fact that renewable energies are dispersed to the network. Consequently their integration into the grid comes at a cost that can significantly vary from one place to another. Finally, different rules exist for power curtailment of renewable energies and how they can be compensated for the corresponding loss of revenue (NREL/Renewable integration 2012). Renewable energies thus face a number of opportunities and constraints which operators must cope with in order to maximize their revenues.

In addition, the fast increase in the share of variable renewable energies (in particular solar) leads to a significant evolution of the competitive landscape in which renewable operators operate. Initially, governments encouraged investors to develop renewable energies by providing guaranteed "feed-in" tariffs to those investing in these types of energies. These prices guaranteed the profitability of the investments no matter what the conditions of operation. These measures created a surge of investments in the power sector. As renewable energies are becoming more competitive, the "feed-in" tariffs are progressively being cancelled to let normal market rules apply. Now, variable renewable energies have very low operation and maintenance costs. The greatest share of their LCOE is made of the initial investment cost. The lower the cost, the lower the cost of electricity and thus the higher the competitiveness. The profitability of the investors relies on the revenue they make. This revenue is obviously a result of the capacity factor, and the ability for operators to maximize the output of their plants. It is also a result of the price at which electricity is sold. When "feed-in" tariffs are no longer applied, variable renewable energies compete in the normal wholesale market. With the increase in the share of renewable energies in the overall mix, the production capacity available at certain times of the day becomes extremely important, pushing prices down at those times. Since the marginal cost of production of renewable energy is null, and operation and maintenance costs very low, prices can reach very low levels at times when these sources of energy produce the most. This leads to a situation where renewable operators' revenues tend to drop since they produce only when prices are low (Solar Asset Management 2015). It thus becomes critical for renewable operators to be able to adjust their production to times where prices are higher, or to participate in market mechanisms (balancing services, operational reserves, etc.) where they can actually get more revenue for their production potential.

To summarize, renewable operators face a situation where they must meet with more demanding requirements in terms of grid connection and look for new ways to maximize their revenue against a backdrop of more volatile wholesale prices and an ever increasingly complex market structure. They must thus find a higher level of flexibility to maintain profitability. The digitization of operations is the main way for them to actually reach the level of flexibility they require.

At the level of the wind or the solar farm, the control of the power plant requires to be digitized to enable real-time control of production. Power Plant Controllers (PPC) include all the necessary algorithms to balance power and reactive power outputs to the grid to both meet the grid codes requirements as well as optimize the output to economic constraints driven by the commitments taken (IEC/Renewable integration 2012). Most renewable operators also operate several farms in various locations. The consolidation of these farms' outputs helps to better balance production, depending on the time and weather conditions. The actual output of a wind or a solar farm at a certain location can indeed vary greatly from that at another location. Consolidating the outputs from geographically distributed can help operators enhance flexibility. Renewable Control Centers (RCCs) are built for that purpose. Such control centers offer benefits to renewable operators. These benefits include optimizing production output and market participation, and better incident management, and lower maintenance costs (Iberdrola/RCC 2016). Some TSOs have also developed such types of control centers to more accurately predict the actual renewable production on the grid and therefore anticipate the measures to be taken to balance the grid (Red Electrica/RCC 2016). RCCs have a multitude of applications. Using advanced weather forecasting technologies, they help provide a more accurate forecast of output from different farms. When connected to trading systems, they can be used to optimize the trade positions that the operator is taking on the wholesale market. With the development of "intra-day" markets, these systems can operate close to real time, maximizing revenue for renewable operators. They also help manage better reactions to incidents by monitoring in real time the status of the farms. Finally, RCCs enable optimized management of assets that are traditionally geographically dispersed by centralizing asset management requirements and workforce dispatch.

The past few years have also seen the development of Virtual Power Plants (VPP), which go one step further in terms of consolidation and balance of renewable energies (Fraunhofer/VPP 2013). VPPs can help better balance the load and production output of renewable energies. They bring a higher level of predictability of production output since they consolidate a vast array of renewable energies and other power sources (wind, solar, biomass, CHP, storage, etc.) and actionable demand. They help operators to optimize their price positioning in the market and maximize their revenue. Finally, they help support grid balance by participating in the reserves markets (U.S. DOE/Ventyx 2010; Caltech 2012). VPPs are traditionally made of real-time monitoring and control of renewable farms. The VPP module corresponds to a Distributed Energy Management System (DEMS), which optimizes in real time the production and scheduling of various renewable plants and loads. The VPP is connected to energy trading, billing services, contract management, and network operation management systems (Renewable Energy World 2013). VPP operators can interact commercially with various renewable plants as well as consumers which are connected to the VPP through a commercial demand response program. VPPs are also a powerful tool to bridge financial and physical flows. Financial flows correspond to the retail and purchase of electricity. These flows are governed by market rules, and theoretically any buyer can purchase electricity from any seller. Physical flows are more complex as energy cannot simply be distributed from one point to another in a specific and discriminated manner. Any load which connects to the grid calls for the demand it needs. Supply sources produce up to what is contracted, and actually to the extent that the sum of the power sources connected on the grid equals the total demand on the network. There are thus up and down variations between what they actually produce and what is contracted, because of evolutions of the demand. VPPs match demand with generation. There are "physical" VPPs that match loads and supply in a physical manner. This concept is very close to a "microgrid". Technically, the "physical" VPP connects to the grid and drives balance between load and supply. There are also "financial" VPPs that associate loads with supply in a financial manner. Here, the purpose is more for a renewable operator to consolidate financial contracts and optimize its own output to the grid in order to meet regulatory requirements as well as optimize its revenues (U.S. DOE/Ventyx 2010). The VPP is thus an interesting concept which is expected to develop fast. The current market for VPPs shall indeed be multiplied by four in the coming 5 years (Renewable Energy World 2013).

In conclusion, renewable operators face an increasingly difficult equation to solve with, on one hand, more regulatory constraints and, on the other hand, a need to optimize their revenue in a difficult wholesale market where renewable electricity prices tend to drop at those times renewable energies are being produced. Digitization is a strong enabler which, through the deployment of power plant control systems, renewable control centers, and virtual power plants, can help renewable operators meet targets and maximize benefits.

5.3 Reaching New Levels of Flexibility with Active Energy Management

Flexibility will be the core of electricity networks of tomorrow. As the load varies during the day, electricity networks have developed over the years to meet an "always on" paradigm (Energy Efficiency Markets 2015). Generation capacity was built to accommodate the highest possible load, and electricity networks designed to supply the maximum possible load throughout the grid. Consequently, the overall utilization rate of these infrastructures has remained quite low. Since the load varied during the day, the supply and distribution of energy was designed for flexibility. Networks and their management systems were designed to cope with these variations as well as the brutal adjustments which could occur following an incident on the network itself or on the supply capacity. Two factors call for a greater need for flexibility. First, the unbundling of electricity markets in mature economies yielded many new actors on the network. Since each looks after optimizing its revenue, there has been an increasing requirement for each of them to build the necessary flexibility to adjust to any change in the market and distribution conditions. Second, the increasing penetration of variable renewable energies leads to a non-fully predictable supply of energy on the network. This variation of supply on the network introduces a new degree of variability to the complex equation for balancing supply and demand in real time. As already explained, these multiple real-time variations lead to an increase in balancing and ancillary services, as well as additional grid costs. In other words, these variations result in inefficiency in the way the network is managed and balanced. Flexibility in supply and demand is thus compulsory to mitigate the natural inefficiencies that variability creates.

A unique equation lies on the demand side. Traditionally, consumption points have not been instrumented and monitored, and they have been considered inelastic. Now, the penetration of digital technologies is enabling a new discovery of these consumption points and their inherent flexibility and efficiency potential. In addition, the emergence of distributed generation and storage technologies, particularly on the demand side, leads to a complete new paradigm on how to build flexibility at the grid level, and what it means for utilities.

5.3.1 Digitizing the Demand Side

Demand management embraces the flexibility tools which can be used on the demand side. It is the most promising option in the coming years for creating significant flexibility at grid level. Demand management corresponds to the adjustments in the load demand which can be done in real time in order to minimize the call for extra services on the supply side. Typically, it can be used to reduce demand at times of peak consumption in order to limit the startup (and, in the long run, the construction) of peak units. Demand management has been around for decades. In the United States, demand management programs have been established since the 1990s. As of today, they correspond to over 1.2 billion dollars of customer incentives; actual peak demand savings are around 13 GW (EIA/DSM 2016). 54% of those savings come from the industrial sector, 25% from the residential sector, and 21% from the commercial buildings sector (EIA/DSM 2016). France was also one of the frontrunners in demand management. In 2000, it elaborated the EJP ("Effacement Jour de Pointe") which developed into a 6 GW capacity overall (© OECD/IEA, Repower 2016). Now, as the need for flexibility arises in electricity markets, demand management programs are growing in importance. With the unbundling of electricity markets in mature economies, many actors are now able to enter and participate in demand management markets, which remain to be fully organized and structured. A recent study by the Rocky Mountain Institute (2015) has shown that residential demand flexibility in the United States could lead to over 10% of grid investment cost savings. These savings correspond to billions of euros at stake.

Demand management capability can be evaluated as a percentage of the peak load value. This percentage is the result of a complex calculation which includes the fraction of end-user applications which are capable of coping with a demand management program and the potential of demand management as a percentage of their load consumption. Real-time consumption must first be known. As a consequence, smart metering systems are a key enabler of demand management programs. They enable the recording in real time of the actual consumption of final consumer points. However, not all end-user applications are capable of coping with a demand management system. Indeed, for some applications, the demand is extremely inelastic (continuous processes for instance, home electric heating in winter, etc.). As well, control capabilities may not be available for controlling the load in real time. It is generally admitted that around 60% of residential and commercial buildings are amenable to demand management programs in OECD countries, but no more than 27% in China (© OECD/IEA, Smart Grid 2011). Finally, the potential of demand management generally varies between 5% and 20% today (© OECD/IEA, Smart Grid 2011; © OECD/IEA, Customers 2011; © OECD/IEA, Repower 2016). Following full deployment of smart meters in the United States, as well as the necessary incentives to enable a true "demand response" market, McKinsey (2010) estimates that up to 20% of the peak load value could now be saved thanks to demand side management programs. At present, the true potential is generally lower because of the actual deployment status of necessary infrastructures. The realization of this potential would completely offset the natural growth of the peak load demand between 2009 and 2019 (McKinsey 2010), leading the actual peak load value in 2019 to equal that of 2009, despite the general growth in electricity consumption. Figure 5.2 shows the load curve from Spain (Red Electrica 2016) and a theoretical load shifting from the peak time to the time of the lowest point of consumption through a demand management program. In this theoretical example, the peak shaving does not exceed 6%, far from the 20% theoretical identified. The ramp up is reduced significantly from 14 GW down to 9 GW. The minimum load balancing is increased to 24 GW instead of 22 GW. The load being more stable, more continuous power generation can be allowed, which leads to a higher efficiency of the grid overall.

Now, the peak in Spain occurs around 2 pm in summer. In the selected scenario for 2030, the net load curve of Spain experiences a slight drop in the middle of the day, as can be seen in the Fig. 5.3, as well as a higher ramp up averaging 10 GW around the end of the day. The peak of consumption is displaced towards the end of the day. Demand side management programs can help shift load from high peak consumption times to times where renewable energies are plentiful, as can be seen

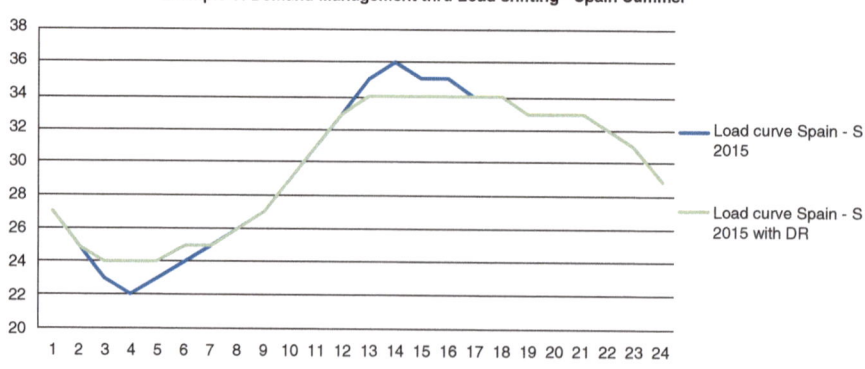

Fig. 5.2 Example of demand management—Spain, summer (Source: Author's own calculation)

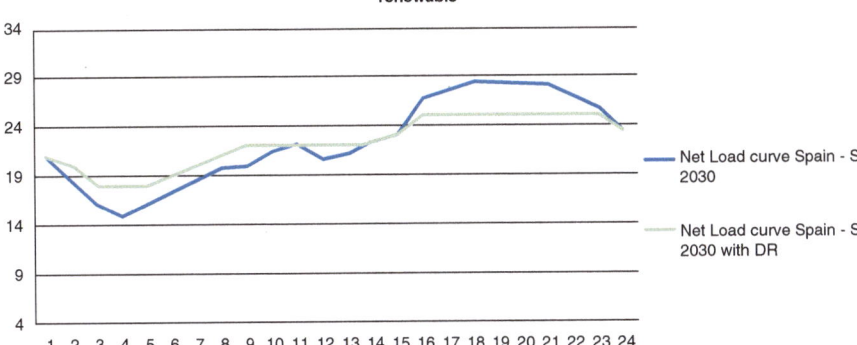

Fig. 5.3 Example of demand management—Spain, summer (with renewable) (Source: Author's own calculation)

on Fig. 5.3. The ramp up is divided by two, down to 3 GW, and the net load curve almost flattened, leading to a much higher level of efficiency of the grid overall. These examples are obviously theoretical. They are meant to exemplify the significant potential of demand side management to optimize grid operation and economics.

There are different forms of demand management programs. These programs can be "characterized as being dispatchable or non-dispatchable" (© OECD/IEA, Customers 2011). This means that some of these programs lead to long-term evolutions of the load curve (energy efficiency incentives for instance), while others correspond to controllable power that can be curtailed (or shifted) on demand. They can also be classified by type of contractual setup. Some of these programs will be based on price incentives, while others will be based on specific programs or "markets" that an aggregator or a retailer will be able to participate in.

Price based demand management programs can be of three types (© OECD/IEA, Customers 2011). Time of Use (TOU) rates can be deployed to favor progressive evolutions of the load curve over time. TOU rates refer to a defined set of retail electricity prices at different times of the day. These prices are defined to reflect the actual average cost of production and delivery during specific time windows. They are typically designed to provide an incentive to consumers to shift load consumption from one part of the day to another in order to reduce the volume of the peak load. The second type of price incentive is Real-Time Pricing systems (RTP). This corresponds to a full reflection of the cost of production and supply of power, with hourly to half-hourly variations during the day. This information is typically available 1 day ahead, similar to power exchanges in the wholesale market. This type of pricing is also an incentive for consumers to adapt their load profile to the evolution of prices. Considering the high degree of variability, it is more importantly an interesting way for retailers to adjust and optimize the power they trade between the wholesale and the retail markets. Typically, RTP rates will lead to a greater need for digital real-time solutions in order to cope with the complex management of the trade between wholesale and retail markets. Finally, Critical Power Pricing (CPP) programs correspond to pricing adjustments related to dispatch balance issues (1 h ahead) and authorization of price spikes in case of scarcity. Scarcity pricing is typically capped. In the case of CPP rates, these prices can go uncapped or capped at a very high level (typically 10,000 USD/MWh, compared to a "normal" 200 USD/ MWh for retail prices for instance). For CPP rates to be efficient, they require a high level of dynamic control capabilities (mostly automated) at the consumer end. They also require the price to be sufficiently high to trigger an incentive. In most mature markets, the general overcapacity of production has so far led CPP programs to be rather inefficient, since prices could not go high enough to trigger a reaction.

Demand management programs can also be part of incentive programs integrated into specific markets that the different actors can participate in. Capacity markets have typically been set up in a number of countries to ensure a reliable energy production mix, guaranteeing a certain capacity is available and dispatchable at any time of the day. They are typically designed to protect some of the conventional generation which is necessary to balance the system at all times. Capacity prices are integrated into the wholesale price and distributed to those of the units which participate in the capacity markets. Capacity can also take the form of "dispatchable" load curtailment. Hence, demand management programs can contribute to capacity markets. They can also play roles in balancing and ancillary services markets. These services are purchased by TSOs to balance the system at all times and are generally purchased very well in advance (months ahead) the day of operation. Similarly to capacity markets, demand side dispatchable load curtailment represents a potential support to balance the grid in real time. Demand bidding programs correspond to bidding load curtailment in the day-ahead wholesale market. Then, demand management programs can participate and contribute to a variety of emergency programs aimed at ensuring the reliability of the network at all times. These programs correspond to contracts where the consumer agrees to reduce its load consumption (either partially through direct load control or fully through curtailment) at times when the operating reserve of the network operator is dropping significantly. All these incentive based programs naturally call for a high level of digitization of both the markets where the actors operate and the physical infrastructure to control loads at times of dispatch.

To sum up, demand management programs are a key resource when it comes to providing flexibility to the grid. Since the marginal cost of production of peak units is very high, a very inelastic demand leads to increased inefficiency and higher prices. A more elastic demand leads to a more efficient grid with lower electricity prices overall. Demand management programs have been in place for decades overall, and have been successfully deployed in a number of countries. Now, the progress of digital technologies enables to go a step further and reach up to 20% of the peak load. Combined with active energy efficiency programs, these technologies enable potentially significant infrastructure savings on the supply side. Smart metering is one of the elements that contribute to a digitized and more flexible demand side. Smart meters are able to collect the data every half an hour instead of once every several months. This data can then be used to profile more accurately the load curve of the consumer in order to suggest the actual efficiency and bill optimization measures that he could undertake. Now, despite the massive deployment of smart metering throughout the world, few raise the concern that these devices might not bring the value they promise. Indeed, most smart metering devices cannot be simply upgraded as technologies evolve. In addition, they do not have control capabilities which are essential in demand management programs, which would therefore require additional infrastructures to be put in place. Some Internet and telecom companies have already started to position themselves in this sector, competing with utilities and retailers and offering a more flexible platform for these programs. Going further, new storage technologies are emerging. Together with digital technologies and combined with distributed generation, they could potentially flatten completely the load curve and dramatically simplify grid infrastructures. In this scenario, new aggregators and intermediary companies would appear (and they have started so in the most advanced regions) and, similarly to VPPs, harness the power of demand side flexibility to create new services offerings to both the grid system (generating efficiency at scale), and demand side users (with lower consumption costs). Actually, this flexibility on the demand side is a main source of potential revenue for renewable operators which are also acting as retailers of electricity, a model frequently deployed across the world. These companies, through digital technologies, manage significant flexibility through complex contract relationships in both the wholesale and retail markets. In such cases, the Virtual Power Plant also integrates flexible demand.

5.3.2   The Fantastic Potential of Energy Storage

Traditionally, the world of electricity has been built on the principle that, on one hand, electricity was a critical commodity which could not afford to be disrupted and, on the other hand, that it could not be stored efficiently. These principles guided investments in the critical infrastructure required to supply and distribute electricity in a real-time and reliable manner. These investments, because of the criticality of the supply, often yielded low asset utilization rates. Priority was given to the continuity of service and the reliability of supply rather than to the economic performance of the investment. Electricity was indeed considered a matter of security and the collateral damage associated with disruption of the supply was estimated to be much greater than the cost of underutilization of the assets. The specifics of electricity supply also yielded a specific market organization, one that is regulated and centralized. The progressive market unbundling as well as the rise in distributed generation have led to a complete new paradigm. Many actors now act in the market with specific interests and capabilities, not all of which have to do with the reliability of supply in real time. In a way, electricity markets have moved in many places from one actor striving to organize balance of supply and demand in real time over a gigantic transmission and distribution network supplying millions of consumer points, to a multitude of actors that somehow agree on how to do the same.

The recent development of energy storage solutions is a disruptive prospect for better balancing of supply and demand and enabling the massive deployment of renewable energies on the grid. Large-scale and distributed power storage capabilities would indeed be able to mitigate variations from renewable energy sources and remove part of the criticality associated with real-time inelastic demand. Energy storage is thus a major challenge and opportunity to ease and enable fully the transition to a new energy system.

5.3.2.1   Different Storage Applications

There are a multitude of applications for storage solutions, as storage enables various critical improvements to grid management as well as power economics (Fig. 5.4). Typically, these applications have different requirements in terms of volume and discharge duration, depending on whether they serve power demand or energy demand. Power applications will require quicker discharge responses, in the range of a few milliseconds to several minutes, while energy applications can require discharge durations up to 10 h. Power applications will typically include quality and reliability support operations, such as frequency regulation and voltage support. Frequency control response time is typically below 1 min, and the discharge duration will range between 1 and 15 min. There can be between 20 and 40 of such operations per day. Voltage support response time typically ranges between 100 ms and 1 s. The discharge duration will not exceed 1 min in general, and these operations can happen up to 100 times a day. Another application associated with quality and reliability applications involves operational reserves management. The grid indeed operates

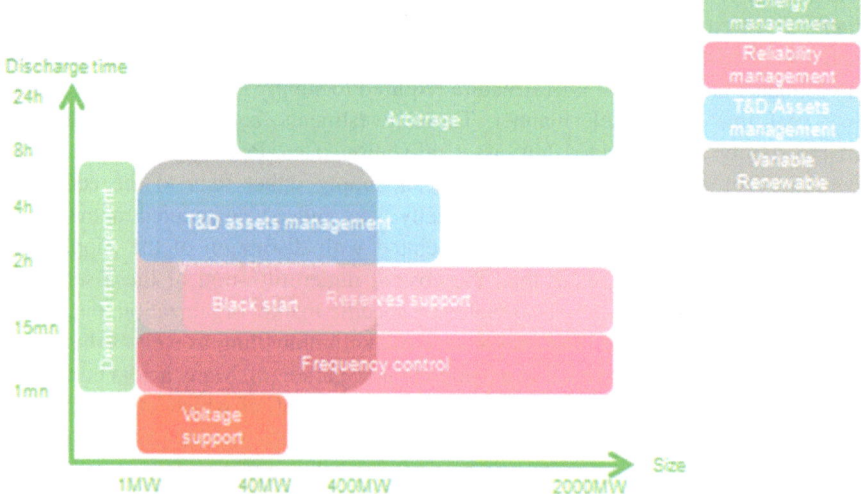

Fig. 5.4   Storage applications (Elsevier 2012; © OECD/IEA, Storage 2014)

always with a certain amount of operating reserves that can be used in case of incident on the supply side. Storage solutions can contribute to the reserves. The response time corresponds to the secondary control of the grid (after frequency and voltage primary controls have reacted to the incident), and therefore ranges below 15 min. The discharge duration can typically range between 15 min and 2 h, and it will happen from zero up to a couple of times a day. A final application of power storage solutions concerns black start support. This corresponds to supporting the progressive ramp up and restart of the full network. Response time does not exceed 1 h. Here, discharge duration can reach up to 4 h, and this is expected to happen less than once a year (Elsevier 2012; © OECD/IEA, Storage 2014). Energy applications are often associated with economics. Demand management applications correspond to shifting loads from periods of high consumption to lower consumption times, thus reducing the peak. Storage is a powerful enabler of demand management programs as it enables the shifting of load from the grid without actually shutting power down on the demand side. The response time of such applications typically range below 15 min. Discharge duration can range between minutes and hours, depending on the specifics of the demand management program. This type of operations can happen several times during the day. Arbitrage is another application which concerns more power suppliers. This consists of storing energy during low-price periods when there is an excess of power supply available and then reselling it during high-price periods. This is particularly an opportunity for renewable operators (in particular solar). With the rise of photovoltaic solar energy, the price at times when the sun shines tends to drop, leading to a loss of revenue for solar operators. Storage provides the opportunity to store this energy and resell it when the demand increases and when the power supply capacity drops, for example, during peak load in the evening. The response time typically is above 1 h and fairly predictable. The discharge duration requirements can reach several hours. Another energy-related application of storage solutions corresponds to transmission and distribution asset management. Storage can help mitigate localized congestion issues. It can also help, on a longer term, to defer infrastructure investments by relying on local storage solutions rather than investing in grid reinforcement (Palmer 2013). This is particularly relevant in remote radial networks, where the cost of the grid reinforcement can be very high for a relatively low utilization rate. Local storage solutions at the end of the line store the energy supplied by the weak infrastructure at times of low consumption, and prevent the demand from the line to exceed its nominal capability at times of high consumption, preventing either congestion of the line or additional grid investments. The response time of such storage applications is typically above 1 h. The discharge duration corresponds to the times of high consumption and therefore can reach several hours. The last possible application of storage solutions concerns directly variable renewable energies. Wind and photovoltaic solar farms lack predictability. Storage at the border of the farm can help mitigate this lack of predictability by artificially de-rating the farm so that a portion of the power generated can be stored. The storage solution can then cope with variations of the power generated to provide a constant and predictable supply of energy at the border of the farm (Elsevier 2012; © OECD/IEA, Storage 2014). This type of solution will become compulsory in all countries where regulators impose on renewable operators constraints on their contribution to the grid balance.

Applications of storage solutions to grid management are thus numerous and can possibly help solve those issues outlined earlier in this book. Notably, a significant deployment of low-power storage solutions in the distribution network (residential, commercial buildings), could help in the development of a coherent demand management program at a much higher scale than those developed today. Utility-scale storage solutions can then help balance the network, in particular with the high penetration of renewable energies. Finally, storage solutions can also engender better and more fluid operation of the wholesale market through arbitrage. Storage solutions thus represent a formidable opportunity to change the rules of the game and facilitate the transition to a complete new grid, with 100% of renewable energy on the network. One could say that without storage solutions, full convergence towards renewable energies is likely to be impractical, as conventional capacities will still be required at the utility scale to balance the grid. All actors theoretically have an interest to support the deployment of storage solutions. TSOs and DSOs can help better balance the network, in particular with a high penetration of renewable energies. They can also optimize their investments by notably deferring investments which might not be profitable. Renewable operators can better cope with grid codes and the requirements to integrate into the grid. They can also better manage price variations on the wholesale market and optimize their revenues. Finally, consumers can take advantage of demand management programs to optimize their electricity bill by actively engaging the market. Overall, storage is a key enabler of a massive efficiency step for power networks. With the inherent efficiency associated with renewable energies (free fuel), storage offers the perspective of rationalizing significantly supply and network infrastructures. The only limitation to its deployment today is the state of development of the technology (Deloitte/Storage 2015).

5.3.2.2 Different Storage Technologies

Several different storage technologies exist today. However, none has yet reached an appropriate level of maturity for mass commercial deployment. The most common technology for storage is pumped hydroelectricity, in which water dams release water and produce electricity on demand. At times of low consumption, the dam can be filled up by pumping the water downstream of the dam. This technology is mature and already in use (© OECD/IEA, Storage 2014), and can reach above 1 GW of power capacity (Elsevier 2012). Other technologies such as Compressed Air Energy Storage (CAES) and batteries are now reaching a state of commercial deployment although they cannot yet be qualified as mature technologies (Elsevier 2012; © OECD/IEA, Storage 2014). Lead-acid batteries have been around for over a century. More recently, new technologies such as sodium-sulfur and lithium-ion batteries have developed, with higher energy densities and longer discharge times (Lazard/Storage 2015). Other technologies such as flow batteries, flywheels, supercapacitors, superconducting magnetic energy storage (SMES) and hydrogen

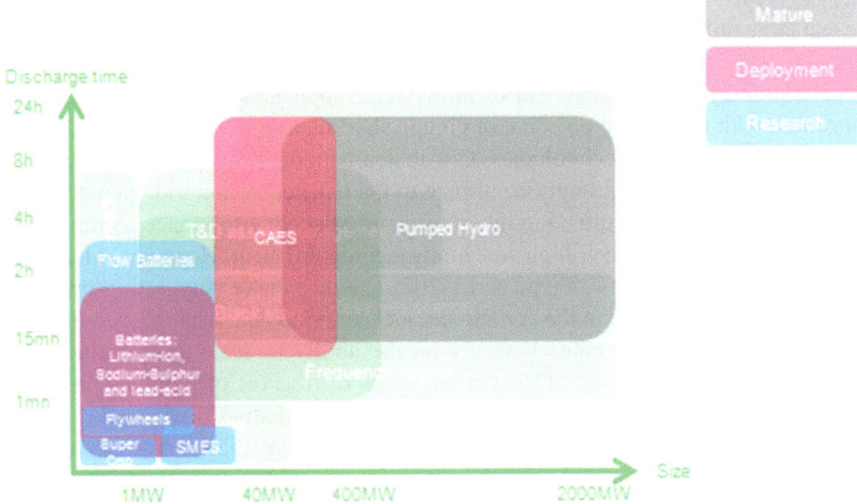

Fig. 5.5 Storage technologies (Elsevier 2012; © OECD/IEA, Storage 2014)

are still being researched (© OECD/IEA, Storage 2014). Figure 5.5 projects the various technologies available or being researched on the actual map of applications identified for storage solutions. Large-scale "bulk" energy applications are well served by pumped hydro and CAES. While pumped hydro is a mature technology, it has limited potential due to its actual use. Indeed, 68% of the base capacity of hydropower worldwide has already been equipped with water dams (Petit 2017). Consequently, the increased need for storage can barely be met with standard pumped hydro technologies. CAESs represent the most interesting alternative to pumped hydro, though it is still in the early deployment stage. Utility-scale storage applications will require large air reservoirs to be able to generate the actual amount of power required. There is a lack of analysis of the full potential of CAES installations worldwide at this stage, and this limits its deployment. There is also uncertainty as to how far the grid could be supported by large storage solutions in a scenario with high penetration of renewable energies. More compact solutions are available at the distribution level. There, batteries predominate. Flow batteries offer a very interesting potential since they can be recharged around 5000—10,000 times, as opposed to traditional batteries which can at best be recharged 1000 times (Naam 2015). The consequence is that, provided they get to mature commercial deployment, the corresponding cost of electricity is theoretically five to ten times lower than traditional batteries. In addition, they offer longer discharge time. These technologies are typically foreseen to be integrated at the demand side and provide power and energy support locally. Batteries offer a great opportunity for demand management programs, and are generally preferred for microgrids. Nevertheless, their cost has so far limited their deployment. Flywheels, supercapacitors and SMES are used for short-term power applications which require rapid response times since their energy density is low. They are being researched today but could possibly complement to an extent large utility-scale storage technologies, notably for power-related applications such as voltage support.

In summary, the development of utility-scale applications has so far remained uncertain with the limited potential of pumped hydro and the lack of analysis on the full potential of CAES technologies. Distributed batteries can enable demand management programs and strongly contribute to the balance of the grid. New technologies based on sodium-sulfur and lithium-ion offer better performances compared to lead-acid batteries, but their high cost limits their deployment. However, it has been observed that they follow a typical 15–20% learning curve, suggesting that they could become affordable in the coming decades, provided the quantities continue to ramp up. A question remains however on the ability to scale up quantities in a sustainable manner because of the current status of reserves of lithium, cobalt and nickel (Greentech Media/Lithium 2015; Greentech Media/Cobalt Nickel 2015) which, according to the experts, would not offer more than a few decades of production (at current pace). Finally, emerging technologies such as flywheels, supercapacitors or SMES can support locally some important power applications such as voltage support.

A variety of studies have been made on the balance between costs and benefits of various storage solutions (Elsevier 2012; Deloitte/Storage 2015). The interest of a given storage technology thus depends on what application it is used for. Typically, arbitrage applications rely essentially on pumped hydro and CAES systems. CAES seems to be the most interesting option from a cost/benefit standpoint. In the future, Deloitte (2015) considers flow batteries and hydrogen solutions could also become interesting options. T&D asset management essentially relies on CAES solutions today. However, flow batteries, SMES and supercapacitors (for sub-second response times) could also become interesting options as they mature. Variable renewable energies integration also relies on CAES today. However, batteries (in particular lithium-ion) seem to be the most promising solution in the future, especially for small distributed applications. Demand side applications essentially rely on batteries (lead acid, lithium-ion and, in future, flow batteries). Lithium-ion and flow batteries show the most promise. Finally, ancillary services such as voltage support are best served today by batteries such as lead-acid and lithium-ion. In the future, SMES and supercapacitors will represent an interesting opportunity for specific applications. In the end, storage solutions have a spectacular potential of development as a very economical solution for many grid services, as opposed to other traditional options available. With the sharp expected decrease of costs for storage technologies (in particular distributed) in the coming years, storage solutions could then progressively compete in the bulk energy market, leading to a complete disruption of electricity markets.

Going further, thermal storage solutions are generally more advanced than power storage. Most thermal storage technologies are already mature (© OECD/IEA, Storage 2014). Underground thermal storage, pit storage, and residential hot water storage are already widely deployed. Thermal storage can be an interesting complement, not to produce electricity but to substitute to a certain extent power demand. It can notably be used to provide space and water heating, and can serve as a substitute when these applications are served by electricity.

5.3.2.3   Storage Economics and Forecasts

The equivalent LCOE for storage is slightly more complicated to evaluate than for basic generation technologies. The costs of storage solutions need to be included into the overall costs of ownership which determine the LCOE. Capital costs (as well as operation and maintenance costs) need to be broken down, taking into account the number of discharge cycles that a storage solution (a battery typically) can experience during its lifetime. This number of cycles will depend on the depth of discharge (DoD) which can vary depending on the application of the storage solution—the deeper the discharge, the lower the number of available cycles. In addition, the efficiency of the round trip (charge and discharge) needs also to be taken into account. It will vary depending on the solution and the specificities of the application. Taking all this into account, a Levelized Cost of Storage can be evaluated. The cost of charging (electricity costs to charge the storage solution) then needs to be added, which will depend on the type of charging (rooftop solar or retail electricity for distributed applications, wholesale electricity for utility applications, etc.). The equivalent LCOE of a storage solution can then be calculated.

The LCOE for storage solutions can be retrieved from various sources. These costs obviously vary from one type of storage to another, with significant differences between minimum and maximum costs. These costs vary significantly from one source of information to another (OECD IEA, Costs 2015; © OECD/IEA, Storage 2014; Lazard/Storage 2015). Figure 5.6 demonstrates this complexity. While pumped hydroelectric storage and CAES seem to have a fairly stable min/max overall cost across sources, there are significant differences for batteries. The International Energy Agency (2015) typically shows for lithium-ion important gaps between a minimum cost of 250 USD/MWh and a maximum cost ten times

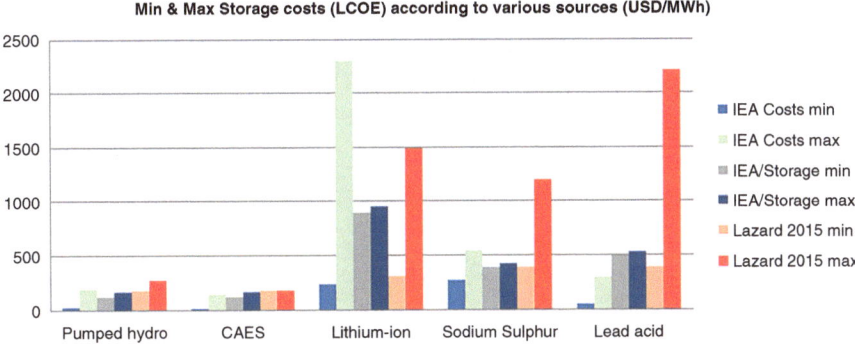

Fig. 5.6 Storage costs variation (© OECD/IEA, Costs 2015; © OECD/IEA, Storage 2014; Lazard/Storage 2015)

higher, while the same agency (2014) shows in another source costs around 900 USD/MWh. For the same type of battery, Lazard (2015) shows an important gap between 300 USD/MWh and 1500 USD/MWh.

These differences can be explained. They actually relate to the typical use cases of these storage technologies. The cost of a storage system is typically lower when the scale is high. Utility-scale applications will tend to have lower LCOE overall because of the scale of power. A certain technology of storage can thus see its LCOE vary significantly, depending on the application. Different use cases also require different discharge and response times, and those technical requirements can also influence cost. The LCOE is notably extremely sensitive to the number of available cycles, which is a function of the depth of discharge. As an example, a specific storage application which would lead to full discharge of the battery once per day would limit the lifespan of the battery to 500–1000 cycles with the current technologies available. Assuming 25% maximum discharge on a daily basis (through a limitation of the output) would extend the lifespan to up to 3000 cycles, thus dividing the LCOE by up to six (Solar Choice 2015)! Many applications thus exist for storage, all with different requirements and consequently different associated LCOEs.

Lazard (2015) has best described these particularities by associating a technology with different use cases and analyzing the variations inside each use case (Fig. 5.7). Three main use cases could be identified. Utility-scale applications cover all applications where storage is directly connected to the grid as a reserve or frequency/voltage support system. PV integration corresponds to applications where storage is deployed at the border of the photovoltaic solar farm, with the objective of stabilizing the grid output. Distributed applications correspond to commercial and residential applications. Typically, distributed applications are more costly than utility-scale or PV integration applications. Within a given technology, there are still important

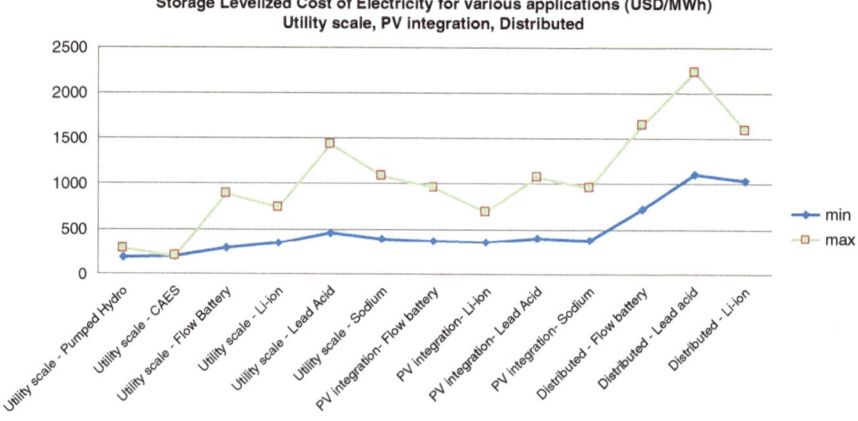

Fig. 5.7 Storage costs variation for various applications (Lazard/Storage 2015)

differences which relate to the various data from manufacturer and the variety of assumptions taken to calculate the cost of investment, operation, tax, etc.

As an overall summary, the average cost of storage in utility-scale applications, including renewable farms (PV integration), range between 200 USD/MWh and 500 USD/MWh, while the cost for distributed storage is much higher, in the range of 1000 USD/MWh–1500 USD/MWh. These figures correspond to one of the most comprehensive studies on storage costs available. They nevertheless need to be taken with caution as a number of assumptions have been made in the use cases (typically the number of cycles and the depth of discharge of the applications) which could differ in other sources.

Going deeper, these storage costs can be broken down by the element which impacts the LCOE the most (Fig. 5.8). The cost of capital is a very significant portion of the total cost of storage. It averages 50–60% of the total cost. Operation and maintenance costs remain low, in particular for utility-scale applications. They can represent up to 13% for distributed applications, but no more than 5–10% for utility-scale applications. Finally, the cost of charging represents a significant portion of the total cost since the corresponding electricity needs to be produced. Cost of charging is naturally higher in distributed applications since the commodity needs to be supplied all the way through the distribution network (except when associated with a rooftop solar system).

These costs are thus very sensitive to the cost of capital and the cost of charging. The cost of capital is essentially related to the technology. These technologies follow a traditional learning curve. The Electric Power Research Institute (EPRI) has evaluated that the costs of lithium-ion batteries drop by 15% per doubling of volume. Bloomberg has estimated this ratio to be around 21%, very similar to photovoltaic solar modules. Naam (2015) has estimated that the associated cost of storage electricity (without charging costs) could reach 130 USD/MWh for 320 GWh (or around 100 GW of capacity if assuming a 3:1 ratio for energy and capacity)

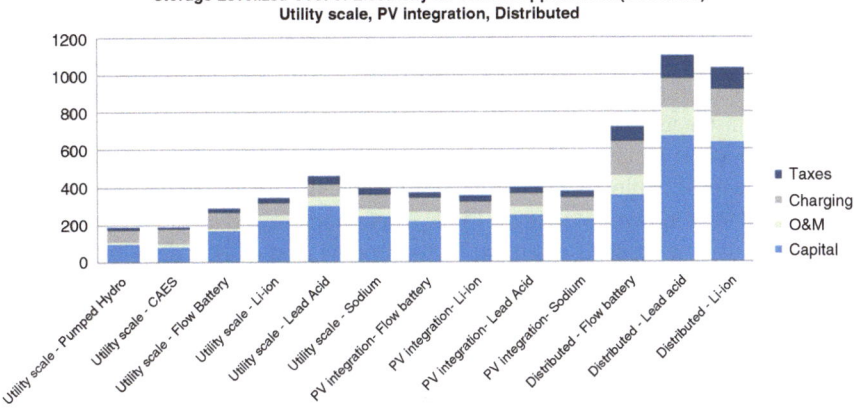

Fig. 5.8   Storage costs variation for various applications (Lazard/Storage 2015)

of batteries deployed (by 2020), and could go as low as 50 USD/MWh for 5000 GWh (or 1600 GW) of storage installed (by 2030). He estimated that this cost would eventually stabilize at around 30 USD/MWh. These projections are globally consistent with a recent report from Navigant (2016) on the deployment of storage worldwide, as well as various references (Energy Manager Today 2014; CleanTechnica/Storage 2015; pv magazine 2014). The scale of deployment of storage technologies will thus be a key factor for its price competitiveness, as for photovoltaic solar technologies.

The cost of charging can also vary significantly depending on how the storage system is being charged, and by which type of energy. Theoretically, the cost of charging of a battery in a residential home equipped with local rooftop solar solution corresponds to the cost of the solar solution itself. In case of excess solar production, and in the absence of "net metering" (resell electricity to the grid), the cost of charging could be considered as null, as the marginal cost of production is zero. It can also correspond to the cost of charging of base load conventional generators running continuously (for instance at times of low consumption. Costs of such generators are generally low, at around 50 USD/MWh. Different use cases thus yield different LCOEs for stored electricity.

The first step in the deployment of storage solutions involves the substitution of traditional applications. Storage solutions can indeed be used to substitute "peakers", generators which are only used to meet peak demand and which cost of electricity is generally well above 200 USD/MWh (wholesale price, not distributed), with the exception of hydroelectric power. The corresponding retail electricity price could reach between 400 USD/MWh and 600 USD/MWh. Batteries can be charged with conventional generation (at low consumption times) or variable renewable energies (at any time they produce, in case of excess of production). With a levelized cost of charging between 0 USD/MWh and 50 USD/MWh, the cost of the battery should range below 350 USD/MWh to compete with "peakers". Storage would then be successfully used to substitute polluting and expensive "peakers". This is already very close to the actual cost of electricity of a number of storage solutions. Storage solutions also contribute to a multitude of applications, such as T&D asset optimization, variable renewable integration, as well as ancillary services such as frequency control, reserve support and voltage control. There, the actual cost of running these applications is much higher than pure bulk energy cost. Elsevier (2012) demonstrated for instance that batteries have already reached for such applications (and for some time) a cost which makes the investment profitable.

With the mass deployment of storage solutions, the associated cost of electricity is expected to go down following a 15–21% learning curve (Naam 2015). A second step could thus be reached when the corresponding cost of electricity would reach parity with bulk energy supply. This will obviously happen first in the retail market, where the cost of electricity is much higher than in the wholesale market (two to three times). Another reason why this step would first be reached at the distribution level is that this is where the scale (and cost) effect of the mass deployment of storage solutions would be the most effective. Charging would then be essentially provided by variable renewable energies since theoretically conventional generation could then be substituted economically. This requires renewable LCOEs to go down sufficiently and the cost of capital of storage technologies to drop as well. Assuming a fixed retail market price of 200 USD/MWh on average (typical in Europe), and the cost of distributed generation (rooftop solar) dropping between 50 USD/MWh and 100 USD/MWh by 2030 (already reached in a number of examples in 2017), storage costs would have to reach between 100 USD/MWh and 150 USD/MWh, which seems consistent with current projections. Obviously, the lower the retail price, the more difficult the economic equation. With a current retail price of around 120 USD/ MWh in North America, storage would have to drop to 70 USD/MWh, assuming distributed solar costs drop to 50 USD/MWh. Reversely, with a retail price of around 330 USD/MWh in Germany, and assuming a progressive reduction of photovoltaic solar costs to 100 USD/MWh, storage would only have to reach 230 USD/MWh to make combined distributed solar and storage solutions competitive with regards to traditional grid costs. It seems thus that combined storage and solar solutions will first become profitable in all regions where retail prices are traditionally high (Europe, OECD Asia), and that they could do so in the coming decade. They would reach parity with bulk energy in other countries eventually. A lot will also depend on the evolution of retail prices, which are on an upward trend everywhere. Where competitive, these solutions would likely take a significant share of electricity markets, thus increasing load defection from the traditional grid.

To conclude, the storage economic equation must first make sense for utilities before it can actually be further adopted as an individual retail load defection solution. This is what the Brattle Group (2014) has clearly demonstrated with the Texas example in a study commissioned by Oncor. It showed that above one billion dollars could be saved in different benefits, in particular deferred investments which would represent over 250 MUSD of net savings (compared to the necessary investment in grid storage solutions). If we project ourselves to 2030, 5000 GWh of storage energy would represent around one and a half hours of the world peak's consumption. This massive amount of production capability could completely modify the shape of the load curve by taming the peak significantly. If we assume for instance that 20% of the maximum load would need to be erased from the grid at times of peak for a period of 3 h on average, in order to flatten the load curve, the amount of storage necessary to do so would not exceed 2000 GWh. Storage thus first represents a disruptive potential in terms of flexible operations of the grid, and it can do so already. As costs fall further in the coming decades, storage will progressively become one of the cornerstones of the new decentralized architecture of power generation.

5.3.3 Electric Vehicles

Flexibility is the main driver of innovation for grid systems of the future. With inelastic demand and variable supply, it has become paramount to balance efficiently one and the other in real-time. Now, demand is also expected to evolve. The recent emergence of electric vehicles is one critical example. Electric engines have efficiency ratios which top 90%, as opposed to traditional combustion engines, which have efficiency ratios below 30%. Electricity supply is traditionally very inefficient, yet the move to renewable energies (in particular distributed energy) has the potential to significantly improve the yield of the power system overall, making the electric engine end-to-end system much more efficient than traditional combustion engines systems.

Electrical vehicles thus represent a fantastic potential for substitution of fossil fuels. Oil consumption for transport represents indeed around 25% of global final energy consumption (© OECD/IEA, WEO 2012). 52% of this energy is devoted to light-vehicle transportation. Out of this, 84% correspond to small-distance travel, typically intra-urban transport. Electric vehicles represent a fantastic alternative to traditional combustion engines for such type of travel. In this chapter, we have not considered additional applications which could emerge in the coming years (heavy-load transportation, marine, etc.). Electric vehicles are recharged by electric power. This represents an equivalent of 8000 TWh of final energy (as of today) switched from fossil fuels to electric power. This however does not mean that the additional electricity consumption required by electric cars would amount up to 8000 TWh, since the efficiency of electric and gasoline cars differ. This thus needs to be estimated. The specificities of regions need to be taken into account too, since there are a great variety of situations.

The share of light-vehicle transportation varies significantly from one country to another. It reaches up to 79% of transport in North America, but barely tops 7% in India. We estimate that the ratios in mature economies should remain stable in the coming decades, while they should strongly increase in new economies with the accelerated economic development of these regions. We have assumed that it would reach 40% in most new economies by 2050 (Fig. 5.9).

On average, 84% of light-vehicle travel is for small distances. The percentage varies from one country to another, from as low as 76% in China to up to 94% in Asia. We have estimated that these figures would remain stable over the years because they correspond to travel habits (Fig. 5.10).

There are today around 1.7 billion cars registered (WHO 2016), about one for every four people on the planet (26%). This ratio shows great variation from one

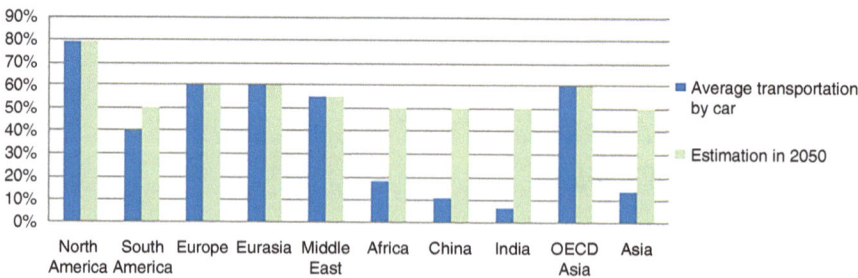

Fig. 5.9 Share of light vehicle transportation (© OECD/IEA, WEO 2012)

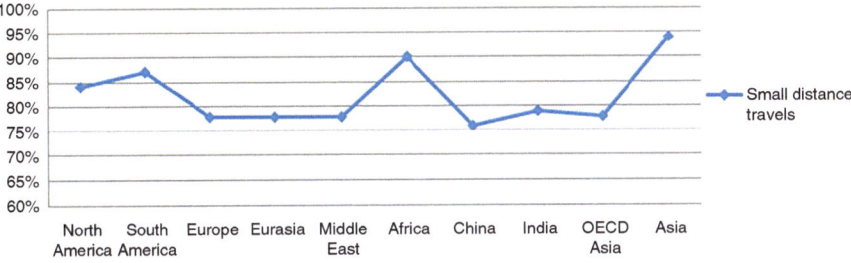

Fig. 5.10 Share of small distance travels (© OECD/IEA, WEO 2012)

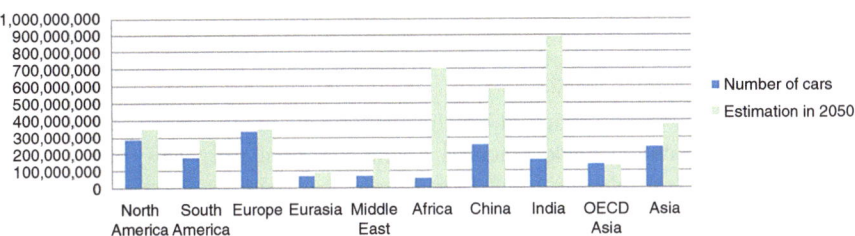

Fig. 5.11 Projection of number of cars worldwide (WHO 2016)

country to another, with around 82% for North America and 6% in Africa. We have as well estimated how this ratio could evolve in the coming decades, assuming stability in mature economies, and steep growth (40% on average) in new economies. The overall worldwide ratio of cars per habitant would then be lifted up from 26% today to 45% (Fig. 5.11).

The mobility of individuals is expected to evolve. On average, people in mature economies travel around 17,000 km per year with their car, against 5000 in new economies. These ratios are set to increase to around 20,000 km per year in mature economies and 12,000 km in new economies (Geohive 2014).

The electric consumption of cars corresponds to around 0.2 kWh/km (FS-UNEP 2016). From this, it can be derived by applying the estimates above that the full substitution of combustion engines by electric cars for small-distance transportation would lead to a certain amount of electricity consumption required. The current full potential corresponds to around 1660 TWh per year, and could reach 5000 TWh in the coming decades (Fig. 5.12).

It is interesting to note that the full substitution today of combustion engines by electric cars would lead to a saving in energy consumption. 8000 TWh would be saved, while 1660 TWh would be required. The saving is slightly lower if we consider electric power to be supplied by conventional generation, because of the efficiency ratio of conventional power plants. In this case, the actual "primary" energy consumed by electric cars would correspond to around 5000 TWh because typically one unit of power consumed corresponds to around three units of fossil

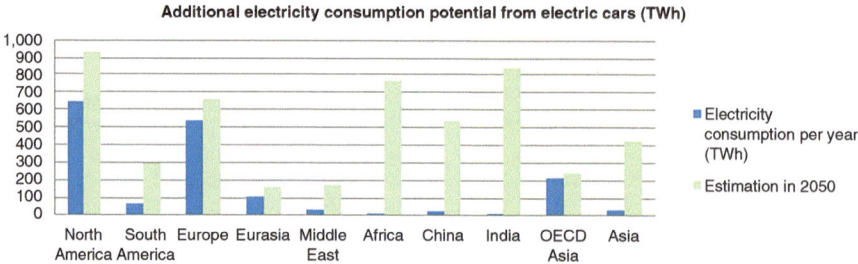

Fig. 5.12 Additional electricity consumption potential from electric cars (Source: Author's own calculation)

primary energy. Still, there would be savings. With electricity being produced by renewable energies however, this inefficiency yield would not apply, and the saving would be more significant.

With a worldwide consumption which would reach around 30,000 TWh per year by 2035, the electric vehicle potential would represent over 15% of electricity consumption increase then. Current forecasts show a steep development of electric cars. The number of electric vehicles could reach six millions by 2020 (FS-UNEP 2016), and up to 25% of the worldwide fleet by 2040 (BNEF/Electric vehicles 2016).

The main issue with electric cars is not so much the additional electricity consumption, but more the actual power capacity it would require in an uncontrolled environment (Greentech Media/EV 2012). Indeed, it is most likely that, without any control, most car batteries would be recharged at night at the end of the day. This is typically when the power consumption peak occurs. Car charging could thus increase the peak volume, leading to important new additional peak capacities being required. This additional capacity could reach around 20–50% of the total capacity installed (G4V 2016), a disruption for grid systems. Additionally, the increase in power consumption at peak times could also lead to a greater need for grid reinforcement, necessitating additional investments in lines and substations.

Control strategies must thus be put in place to prevent this effect. There are different approaches to the integration of electric vehicles into the grid (G4V 2016). The "business as usual" approach embeds various strategies, such as enforcing "soft charging" (slow and low-power charging), or use "time of use" tariffs (tariffs which vary depending on the time of the day, typically higher at the time of the peak). A more "pragmatic" approach exists (G4V 2016). It consists of giving the DSO the means to actually control battery recharging in some way, in particular in case of emergency, such as network contingency. The DSO would basically be given the ability through real-time adjustments to perform operations on the charging process, on randomly selected charging stations, or on all of them, depending on the strategy selected. These operations could consist of interrupting the charging process, or changing the speed of recharging (and thus the power associated). Finally, more "advanced" approaches are being studied. These would basically be market based. They would take stock of the fact that electric vehicles' batteries are also in a situation to contribute to the grid balance as simple storage solutions. These market based approaches could (or not) use aggregators to develop larger-scale flexibility solutions for the grid, by adjusting in real time the charging process against a clear retribution scheme. Electric vehicles thus represent an interesting solution for grid flexibility. It could as well become the most significant storage option in the coming years. The integration of their potential into the grid is thus critical and a spectacular opportunity (this becomes a risk if nothing is done) to increase the level of flexibility of the grid, and thus tame investments in grid reinforcement and peak capacity. With the development of renewable energies, electric vehicles also offer an interesting potential as an energy storage solution (Ziolek 2012). Electric vehicles' batteries could indeed help balance out the production from variable renewable energies at all times, as a standard distributed storage solution. Grid for Vehicles (2016) has evaluated that the level of curtailment on renewable power required (in case of high power production from renewable farms) could be divided by eight.

Summing up, electric vehicles represent a potential significant increase in electricity consumption (but overall a spectacular gain in efficiency). They also present a risk for grids in case of absence of control. Indeed, the increase in recharging at times of peak could lead to significant increases in the volume of peak power required, necessitating additional investments in power capacity ("peakers") and grid reinforcement. Now, electric vehicles also offer an opportunity for an increased efficiency of the grid overall. Indeed, batteries can be considered as storage solutions which would be plugged into the grid and available for balancing out power, notably from renewable energies. Such market based "advanced" control solutions are emerging today. At present, the number of electric cars and charging stations has not yet reached a level significant enough to force the market to evolve. This is however undoubtedly coming up, together with the accelerated development of storage solutions.

5.3.4 Telecom and Utility Infrastructures: A Possible Convergence?

The transition to a new market paradigm is offering a number of opportunities at the consumer side. Demand management, distributed energy resources and storage integration are all revolutions which are happening at the residential or commercial level, in the "last mile" where energy is consumed. The emergence of connected technologies offer, as we saw, tremendous opportunities for flexibility and decreased electricity costs over time. There is thus an emerging market opportunity for providing energy management services to the consumer, in order to both relieve the network from excessive constraints as well as to optimize the consumers' bill.

Utilities are obviously uniquely placed to take ownership of these services. Now, different surveys put the capability of utilities to do so in question. A study from IPSOS-Mori (Greentech Media/telecom 2016) evaluates consumers' loyalty towards the energy industry versus the average of different other industries. It shows a slight positive average, with a strong positive result in countries like South Korea, France, Russia, Italy and Saudi Arabia, and a strong negative result in countries such as Germany, Brazil, Australia, the United Kingdom and South Africa. In the United States, 73% of the consumers would consider buying electricity from non-utility companies, and 22% would consider purchasing it from phone or cable companies. This ratio even tops 90% in China and South Korea (Greentech Media/telecom 2011). The capability of utilities to position themselves in these emerging services markets is thus in question in a number of countries, and the consumers themselves, either because they hold a negative opinion of their utility, or simply because they consider these types of services as not naturally coming from utilities, are often tempted to ascribe such types of services to other companies. Customer intimacy is thus a cornerstone of the strategy to capture a share of such markets.

Telecom companies have in many places championed end consumer intimacy for years, providing a set of personalized solutions and services to them. They thus naturally look at these markets as an opportunity for them to expand their portfolio. In Australia, Telstra has already entered this market, providing different services to manage homes, including energy, as part of its "connected homes" strategy (Greentech Media/telecom 2016). Telecom companies are thus progressively engaging this market, and the improvements in digital technologies are a strong enabler. They become competitors of utilities, and utilities are lagging because of their often deficient customer intimacy.

In other countries, recent years have seen a convergence of both industries. NineStart Connect is an emblematic example of such convergence, with a utility company and a telecom company merging to join forces on an overlapping customer base. The new company, a marriage of Central Indiana Power (CIP) and Hancock Telecom Company, has been able to both increase its overall customer base, generate efficiencies in the management of its assets, and develop new services for its customers (T&D World 2011; Greentech Media/telecom 2011). There are indeed a number of advantages on paper from converging both infrastructures. Both industries need to sustain an extended network infrastructure that extends to end consumers. The combined management of both offers significant potential savings in asset management (SmartGridNews 2013). Also, new energy management services require an efficient telecommunication network (inside homes), as well as deep knowledge on how to drive such types of services like demand response, dynamic pricing, etc. (Greentech Media/telecom 2011). The combination of both offers true business perspectives.

In addition, the convergence of data sources also helps develop a new level of aggregated information (multi-facility, multi-utility) which offers expanded insights into end consumer behavior as well as opportunities for controlling the potential flexibility associated with such behavior (Black & Veatch 2016).

In conclusion, with the digital era spreading through homes and consumers, a new spectacular market opportunity is emerging. It will be a highly competitive landscape featuring various industries jostling for customer intimacy. Both utilities and telecom companies are well positioned to capture this market opportunity, but they could also be displaced if they are unable to react fast enough. The convergence of the two industries also creates interesting opportunities to leverage on assets management as well as to combine expertise to develop unique services around the transition to a new grid.

5.3.5 Summary

Digitization is a key enabler of the transition of the energy system to one that is more efficient and sustainable. Digitization will enable DSOs to face the rapid increase in penetration of renewable energies into the network, and maintain reliability of supply as well as an acceptable level of utilization of grid assets. Renewable operators may optimize their revenues thanks to digital technologies, aggregating various renewable farms (and conventional systems) into Virtual Power Plants with a greater level of flexibility, a key success factor in very competitive and increasingly complex markets. The most ample opportunity lies on the demand side. Long considered inelastic, the demand side is being completely reshaped by the growing penetration of decentralized renewable energy, local storage solutions and the progressive emergence of variable loads, such as electric vehicles. Once connected to the rest of the system, this flexible demand side enables a change of paradigm in terms of grid and market operation, enabling further optimization of the efficiency of the system's operation in real time while improving reliability. Going further, the very notion of "demand side" is growingly challenged. If everything now happens there, the traditional separation between "supply" (the world of utilities) and "demand" (the inelastic, invisible consumption points) is fading away. The future grid system is horizontalizing, with the various nodes of supply and consumption interconnected together into a larger network. Not surprisingly, this offers tremendous opportunities for innovation. Traditional incumbents are trying to reposition themselves, but not without struggle. They are at the forefront of this transition. As their daily operations get more complicated than ever, they leverage their traditional access to consumers (if they have retail operations) to rebuild value. New actors emerge as well, such as aggregators trying to capture the value of flexibility by acting as intermediaries to traditional utilities and reaping benefits in the middle. Many startups have been created in the last 5 years with that purpose alongside Internet and telecommunication companies which are starting to take interest in this emerging opportunity. Other startups are even more ambitious; they are considering the opportunity of a totally meshed network of energy, where peer-to-peer exchanges could be organized without recourse to a traditional operator in the middle. Recent technology developments such as blockchain technologies, which have begun to pervade the finance sector for the very same reason, are increasingly being considered. No matter the shape this energy transition will eventually take, or the pace at which it will evolve, it can be safely asserted that the future of energy systems definitely resides on the demand side.

5.4 Different Pathways to a Successful Transition

The transition to a new energy system is in progress everywhere. In the coming decades, the cost of electricity from renewable energies is expected to drop significantly, and become competitive with that from traditional conventional generation solutions. By then, variable renewable sources of energy will dominate. The storage cost of electricity is expected to follow a similar learning curve (in particular batteries), and become affordable. The combination of solar or wind power with storage will thus become competitive, and likely grow to emerge as the bulk of power generation. This solution will be distributed in essence, pervading all residential homes, commercial and industrial buildings. Add to this the fact that utility-scale farms are by nature distributed and it becomes clear that there will arrive a new paradigm governing how energy is produced and distributed. It is only a matter of time.

In 20 years, the new power grid will fundamentally differ from the current landscape which we have known for the last 70 years. The new grid architecture will focus on distributed renewable energy sources of production. A first layer will combine solar (or wind) and storage at the residential, commercial and industrial building level. A greater share of energy will be self-generated. Then, energy sources and consumption points will be aggregated in microgrids at the district or municipal level. These microgrids will connect to a greater interconnected transmission network, which will be tasked with ensuring overall network stability. As well, a significant share of utility-scale distributed energy resources and storage solutions will be consolidated. Conventional power plants, traditionally connected to the transmission network, will serve as capacity to support any incident or temporary lack of output from renewable sources—they will no longer operate continuously on the grid. This network architecture will be mirrored by a similar market architecture. A market for distributed resources will need to emerge to consolidate local distributed energy sources as well as demand management alternatives. Retailers and new players will consolidate these sources and interact in the different markets to share energy, capacity and flexibility services. In general, there will be less and less energy traded (TWh) and more and more power capacity services (GW). On one hand, self-generation will likely take a greater share. On the other hand, the fact that renewable energies come at zero marginal cost will push markets to move away from energy-based (TWh) retribution systems towards more capacity (GW) access retribution systems, similar to what the telecom industry experienced during the first decade of the twenty-first century. The shape of future markets will also likely depend on how the change is driven, as well as the legacy market structure in different countries.

A number of policies either already exist or are in preparation to accelerate and organize this transition (© OECD/IEA, Repower 2016). Carbon regulation can help accelerate the shift to renewable energies, making fossil based technologies less competitive. A greater efficiency of market operations can be achieved going forward with a more dynamic pricing system. Taking stock of the distributed aspect of variable renewable energies, a higher geographical resolution leads as well to a more realistic pricing structure. The capacity market remains to be organized in a number of regions, with capacity prices and pricing for flexibility solutions such as demand response. Finally, more distributed market mechanisms require to be put in place to facilitate the increasing penetration of distributed energy resources as well as consumer engagement.

The pathways to this new model are likely to differ significantly, depending on two main factors: the speed and the extent of change. The faster the change, the more difficult it will be for existing actors to adapt, hence the greater the need for restructuring. Additionally, the larger the change, the more profound will be the need to evolve. The National Renewable Energies Laboratory (2015) has defined four quadrants that describe possible transition pathways (Fig. 5.13). When the change is slow and not profound, existing actors will have to "adapt". When the change is fast but the extent of the change is low, then regulatory authorities will essentially need to "reconstruct" the market. When the change is slow but the extent of the change is profound, current players will need to "evolve". In the last quadrant, when the change happening is both fast and profound, the existing actors will experience a "revolution".

Each geography can be mapped against these scenarios. A specific rating needs though to be given to each region/country in order to provide an accurate assessment of the allocation of countries to a given scenario (Table 5.1). The speed of change is here defined against two criteria: the actual potential for renewable energies (its competitiveness), and the power demand evolution—the more the two factors, the faster the expected speed of change. The extent of change can also be defined against two criteria: the extent to which the current market structure favors the transition, and the opportunity for new entrants to play in the market. Again, the more the two factors, the deeper the changes. Regions have been mapped according to this model. Both Canada and the United States have been split between lagging and advanced

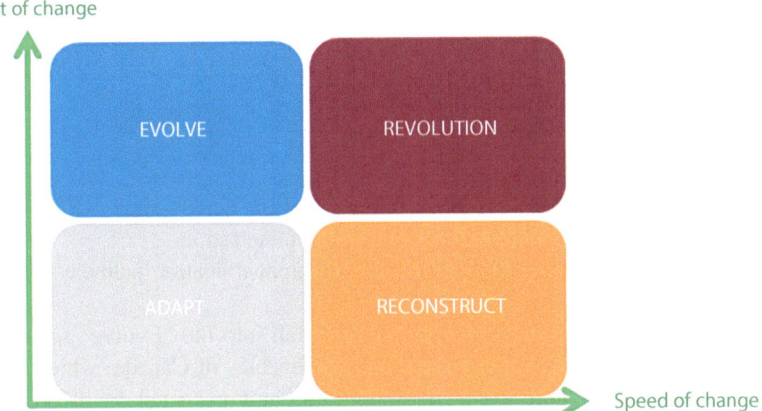

Fig. 5.13 Different pathways for the transition (NREL/Power Future 2015)

Table 5.1 Country evaluation against transition scenarios

Region/Country	Speed: changes in system principles			Extent: changes in competitive landscape		
	Potential for renewable	Power demand evolution	Total	Market structure favoring transition	Opportunity for new entrants	Total
Africa (ex Maghreb and South Africa)	4	3	7	1	3	4
Australia	3	1	4	2	3	5
Canada (adv)	1		2	2	2	4
Canada (lag)	1	1	2	1	1	2
China	2	3	5	2	1	3
Eurasia	1	1	2	1	1	2
France	2	1	3	2	2	4
Germany	3	1	4	2	3	5
India	3	4	7	2	3	5
Italy	3	1	4	2	2	4
Maghreb	4	3	7	1	3	4
Mexico	3	2	5	2	2	4
Middle East	3	2	5	1	2	3
OECD Asia (ex Australia)	1	1	2	2	1	3
Other Europe	2	1	3	2	2	4
Other South America	2	2	4	1	2	3
South Africa	3	2	5	2	3	5
South America (Argentina, Brazil, Chile, Columbia, Ecuador)	2	2	4	2	2	4
South East Asia	3	3	6	1	3	4
Span	3	1	4	2	2	4
UK	2	1	3	2	2	4
US (adv)	3	1	4	3	3	6
US (lag)	2	1	3	1	3	4

Source: Author's own evaluation

states, since there exists a considerable variation across states of both market organization as well as potential for renewable penetration.

These results could be mapped on the different transition pathways (Figs. 5.14 and 5.15).

The analysis shows most OECD countries fall into the "Evolve" scenario, with the exception of Japan, South Korea, and some states of Canada, which fall in the "Adapt" scenario. These countries indeed are not expected to drive important changes because of their existing market structure as well as the limited competitiveness of renewable energies. Overall, the speed of change is not likely to be fast

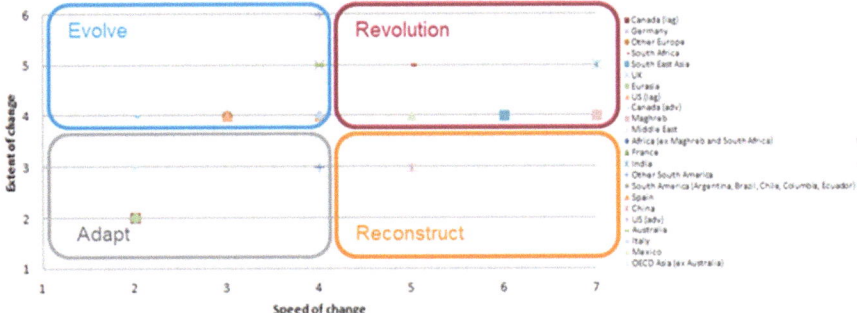

Fig. 5.14 Mapping of countries according to speed and extent of change (Source: Author's own evaluation)

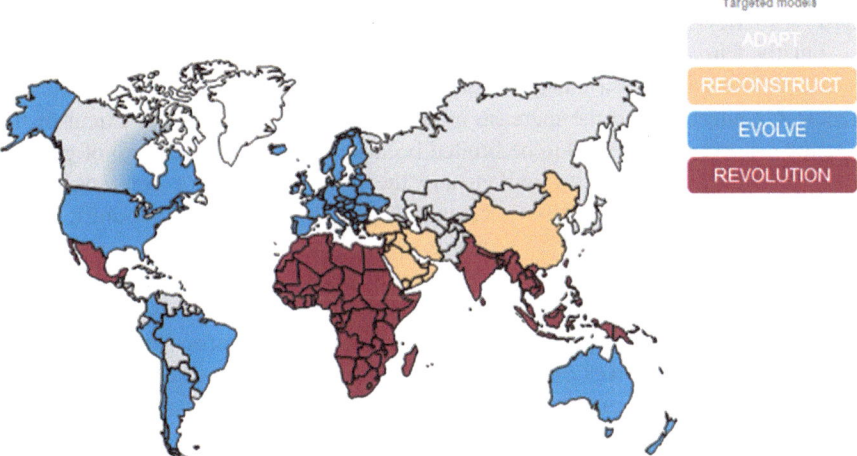

Fig. 5.15 Different pathways for the transition to new energy landscape (Source: Author's own evaluation)

across OECD countries because, despite a variable but definite potential for renewable energies, the growth of electricity consumption is expected to remain low. Obviously, this high-level estimate hides significant disparities—some states or countries are actually closer to a "Revolution" scenario. As previously mentioned, California, Germany and, to a lesser extent, Australia have strong potential for renewable energies and are about to face significant challenges with regards to operating their grid. They could, to a certain extent, be classified under the "Revolution" scenario.

Going further, South America presents significant disparities with Argentina, Brazil, Chile, Columbia and Ecuador being rather advanced in their market organization, and therefore meeting the conditions for an "Evolve" scenario.

Eurasian countries generally fall in the "Adapt" scenario. Most of them operate under a verticalized utility model. In addition, the potential for renewable energies there remains today limited.

A large array of regions, including regions and countries such as Africa, South East Asia, Mexico and India present the conditions for a "Revolution" scenario. These countries indeed have both a strong potential for renewable energies as well as a skyrocketing electricity consumption forecast in the coming decades, leading to significant needs and pressure on the electrical network. In addition, those are the regions with the greatest needs in terms of access to energy. Most of these countries however have very conservative market organizations. For the most part, these power markets are operated by verticalized and regulated utilities. Nevertheless, the relative inadequate balance between a conservative market structure and spectacularly growing needs presents an opportunity for new entrants to bring technologies such as microgrids and capital investment into the landscape. The extent of the change is thus likely to be deeper in those countries, leading overall to a "Revolution" scenario.

Finally, China and Middle East countries are expected to fall in the "Reconstruct" scenario. The speed of the change is there, thanks to both the high competitiveness of renewable energies and the increase in electricity consumption. The extent of the change is however expected to be limited because of the relative control of government organizations over the market, and the limited ability for new entrants to operate in those countries. This is essentially what separates those countries from the ones under the "Revolution" scenario.

Each quadrant contains a number of options on how electricity markets are most likely to transition. In the "Adapt" scenario, verticalized utilities will own the transition. They will progressively adapt to more of a performance based model with concurrent objectives of customer satisfaction, performance of their assets, as well as penetration of renewable energies. This will constitute an important change compared to the traditional model where utilities are guaranteed a retribution for taking care of assets, with less concern over consumers or asset performance. In the "Revolution" scenario, countries are in great need of energy. Electricity consumption increases significantly year after year as the economy develops and the population grows. Vertical structures are often not able to satisfy this growth. In addition, public financing is an issue because of the high interest rates applied to these economies. Governments are thus more likely to prefer private investments (new entrants) to attract more financing, and favor decentralized solutions, which are a more economical alternative than "top-down" grid expansion. This is because a surge of investments in opening markets is more likely to cause both a profound and a fast change. In the "Reconstruct" scenario, utilities will face issues similar to those in the "Revolution" scenario countries. The transition is however expected to be controlled by the incumbent organizations, limiting the possibility for new entrants to operate. The market would then be slowly rebuilt. Finally, the "Evolve" model is looking at a more profound change in the market, with the emergence of distributed generation and microgrids as well as a more distributed market organization. These countries (or states) will reorganize their market faster to unleash the potential of distributed generation and demand management solutions. Most OECD countries are following this path, along with a selected number of countries in South America.

In conclusion, each country will likely follow a different pathway towards the new market model because of the legacy of its market structure as well as the expected evolution of both its consumption and generation mix in the coming years. These elements (legacy organization and asset base, expected renewable penetration rate, expected consumption evolution) will drive the extent, speed and depth of the change. The mapping of countries and their possible transition pathways helps aid the understanding of the dynamics of each country and how the existing actors in the local market will react and participate in this transition. Bottom line, a small share of new economies is expected to slowly transition, maintaining firm control over changes, while others will move in an accelerated manner towards the new model as they have neither a legacy nor a great need for financing and infrastructure. In mature economies, countries will transition to new market models and highly distributed generation, but those changes (including restructuring of the current fleet of power generation units) are not expected to be fast because of the slow consumption evolution and the current lifecycle and competitiveness of the existing generation fleet.

References

© OECD/IEA, Costs. (2015). *Projected costs of electricity.* https://www.iea.org/bookshop/711-Projected_Costs_of_Generating_Electricity

© OECD/IEA, Customers. (2011). *Empowering customer choice in electricity markets.* http://www.iea.org/publications/freepublications/publication/Empower.pdf

© OECD/IEA, Repower. (2016). *Re-powering markets.* http://www.iea.org/publications/freepublications/publication/REPOWERINGMARKETS.pdf

© OECD/IEA, Smart Grid. (2011). *Impact of smart grid technologies on Peak Load to 2050.* https://www.iea.org/publications/freepublications/publication/smart_grid_peak_load.pdf

© OECD/IEA, Storage. (2014). *Technology roadmap—Energy storage.* https://www.iea.org/publications/freepublications/publication/TechnologyRoadmapEnergystorage.pdf

© OECD/IEA, WEO. (2012). http://www.worldenergyoutlook.org/publications/weo-2012/

Black & Veatch. (2016). *Network convergence offers savings and new technologies to utilities.* http://bv.com/Home/news/solutions/Smart-Cities-Telecom/network-convergence-offers-savings-and-new-technologies-to-utilities

BNEF/Electric Vehicles. (2016). *Electric vehicles to be 35% of global new car sales by 2040.* http://about.bnef.com/press-releases/electric-vehicles-to-be-35-of-global-new-car-sales-by-2040/

Brattle Group. (2014). *The value of distributed electricity storage in Texas.* http://www.brattle.com/system/news/pdfs/000/000/749/original/The_Value_of_Distributed_Electricity_Storage_in_Texas.pdf

Caltech. (2012). *Virtual power plants in competitive wholesale electricity markets.* http://resnick.caltech.edu/docs/d-Siemens_Grid3.pdf

CleanTechnica/Storage. (2015). *Distributed energy storage revenue to exceed $16.5 billion by 2024.* http://cleantechnica.com/2015/01/13/distributed-energy-storage-revenue-exceed-16-5-billion-2024/

Deloitte/Storage. (2015). *Energy storage: Tracking the technologies that will transform the power sector.* http://www.ourenergypolicy.org/wp-content/uploads/2016/03/us-er-energy-stor age-tracking-technologies-transform-power-sector.pdf

EIA/DSM. (2016). *Demand response saves electricity during times of high demand.* http://www.eia.gov/todayinenergy/detail.cfm?id=24872

Elsevier, Wang. (2012). *Prospects for renewable energy meeting the challenges of integration with storage.* http://refman.et-model.com/publications/1807

Energy Efficiency Markets. (2015). *The shift to integrated demand-side management programs.* http://energyefficiencymarkets.com/the-shift-to-integrated-demand-side-management-programs/

Energy Manager Today. (2014). *Installed energy storage for the grid to be 20.8 GW by 2024.* http://www.energymanagertoday.com/installed-esgas-20-8-gw-2024-0105018/

Eurosunmed. (2015). *EUROSUNMED workshop: Grid code for renewable energies—Integration in the electric grid.* http://www.eurosunmed.eu/news/eurosunmed-workshop-grid-code-renew able-energies-integration-electric-grid

Fichtner. (2010). *Grid codes for wind power integration in Spain and Germany: Use of incentive payments to encourage grid-friendly wind power plants.* http://www.efchina.org/Attachments/ Report/reports-efchina-20100811-en/Grid%20Codes%20for%20Wind%20Power%20Integra tion%20in%20Spain%20and%20Germany.pdf

Fraunhofer/VPP. (2013). *The virtual power plant stable supply of electricity from renewable energies.* https://www.fraunhofer.de/en/press/research-news/2013/march/the-virtual-power-plant.html

FS-UNEP. (2016). *Frankfurt School UNEP collaborating center on climate and sustainable energy finance.* http://fs-unep-centre.org/sites/default/files/publications/globaltrendsinrenewableenerg yinvestment2016lowres_0.pdf

G4V. (2016). http://www.g4v.eu/downloads.html

Geohive. (2014). http://www.geohive.com/earth/his_history1.aspx

Greentech Media/Cobalt Nickel. (2015). *Why lithium isn't the big worry for lithium-ion batteries.* http://www.greentechmedia.com/articles/read/Why-Lithium-Isnt-the-Big-Worry-for-Li-ion

Greentech Media/EV. (2012). http://www.greentechmedia.com/articles/read/Startup-EV-Grid-to-Pioneer-Vehicle-to-Grid-Technology

Greentech Media/Lithium. (2015). *Is there enough lithium to maintain the growth of the lithium-ion battery market?* http://www.greentechmedia.com/articles/read/Is-There-Enough-Lithium-to-Maintain-the-Growth-of-the-Lithium-Ion-Battery-M

Greentech Media/Telecom. (2011). *The merger of telecom and utilities: Is it the future?* http://www.greentechmedia.com/articles/read/the-merger-of-telecom-and-utility-services-is-it-the-future

Greentech Media/Telecom. (2016). *Will telecom companies kill utilities?* http://www.greentechmedia.com/articles/read/Will-Telecom-Companies-Kill-Utilities

IAM. (2016). *Institute for asset management.* https://theiam.org/knowledge/diagrams

Iberdrola/RCC. (2016). *Renewable energy operation center.* http://www.iberdrola.es/reputation-sustainability/innovation/renewables/renewable-energies-operation-center/

IBM. (2016). http://www-03.ibm.com/software/products/en/ibmmaximoforutilities

IEC/Renewable Integration. (2012). *Grid integration of large-capacity renewable energy sources and use of large-capacity electrical energy storage.* http://www.iec.ch/whitepaper/pdf/iecWP-gridintegrationlargecapacity-LR-en.pdf

Irena/Grid Codes. (2016). *Scaling up variable renewable power: The role of grid codes.* http://www.irena.org/DocumentDownloads/Publications/IRENA_Grid_Codes_2016.pdf

JISEA. (2015). *Joint institute for strategic energy analysis. Renewable electricity: Insights for the coming decade.* http://www.nrel.gov/docs/fy15osti/63604.pdf

Lazard/Storage. (2015). *Levelized cost of storage analysis version 1.0.* https://www.lazard.com/media/2391/lazards-levelized-cost-of-storage-analysis-10.pdf

McKinsey/Digital Utility. (2016). *The digital utility: New opportunities and challenges*. http://www.mckinsey.com/industries/electric-power-and-natural-gas/our-insights/the-digital-utility-new-opportunities-and-challenges

McKinsey/DSM. (2010). *The smart grid and the promise of demand-side management*. https://www.smartgrid.gov/files/The_Smart_Grid_Promise_DemandSide_Management_201003.pdf

Naam, Ramez. (2015). *Why energy storage is about to get big and cheap*. http://rameznaam.com/2015/04/14/energy-storage-about-to-get-big-and-cheap/

Navigant. (2016). *Microgrid development tracker*. http://www.navigantresearch.com/research/microgrid-deployment-tracker-2q14

NREL/Power Future. (2015). *Power systems of the future*. http://www.nrel.gov/docs/fy15osti/62611.pdf

NREL/Renewable Integration. (2012). *Integrating variable renewable energy in electric power markets: Best practices from international experience*. http://www.nrel.gov/docs/fy12osti/53732.pdf

Oracle. (2014). http://www.oracle.com/us/industries/utilities/utilities-work-asset-management-br-046553.pdf

Palmer, G. (2013). *Household solar photovoltaics: Supplier of marginal abatement, or primary source of low-emission power?* http://www.mdpi.com/2071-1050/5/4/1406

Petit, V. (2017). *The energy transition*. Cham: Springer.

pv magazine. (2014). *Forecast 2030: Stored electricity at $0.05/kWh*. http://www.pv-magazine.com/news/details/beitrag/forecast-2030%2D%2Dstored-electricity-at-005-kwh_100016581/#axzz4OfEOzpre

Red Electrica/RCC. (2016). Control center of renewable energies. http://www.ree.es/en/activities/operation-of-the-electricity-system/control-centre-renewable-energies

Renewable Energy World. (2013). *Virtual power plants: A new model for renewables integration*. http://www.renewableenergyworld.com/articles/print/volume-16/issue-5/solar-energy/virtual-power-plants-a-new-model-for-renewables-integration.html

RMI/Flexibility. (2015). *Rocky mountain institute. The economics of demand flexibility*. https://www.rmi.org/insights/reports/economics-demand-flexibility/

Schneider Electric/AMI. (2016). *Smart metering platform for rollout and operation*. http://www.schneider-electric.com/en/product-range/61766-titanium-advanced-meter-infrastructure/

Schneider Electric/SHG. (2016). *Self-healing solution for underground open-loop networks*. http://www.schneider-electric.com/solutions/ww/en/sol/44778368-self-healing-solution-for-under ground-open-loop-networks

Schneider Electric/Utility Solutions. (2016). http://www.schneider-electric.com/b2b/en/solutions/for-business/electric-utilities/address-your-challenge/

Sempra Energy. (2015). *Management system (ADMS)*. http://www.nrel.gov/esi/pdfs/agct_day1_bialek.pdf

SmartGridNews. (2013). *Why electric power and telecom providers need to be talking*. http://www.smartgridnews.com/story/why-electric-power-and-telecom-providers-need-be-talking/2013-08-01

Solar Asset Management. (2015). *Unlocking solar with big data and digitization*. http://www.solarassetmanagementeu.com/new-updates-source/unlocking-pv-solar-with-big-data-and-digitization

Solar Choice. (2015). *Levelised cost of storage: A better way to compare battery value*. http://www.solarchoice.net.au/blog/levelised-cost-of-storage-compare-battery-value

T&D World. (2011). *A new breed of cooperative*. http://tdworld.com/business/new-breed-cooperative

U.S. DOE/ADMS. (2015). *U.S. Department of Energy/Office of electricity delivery and energy reliability. Insights into advanced distribution management systems*. https://www.smartgrid.gov/files/ADMS-Guide_2-11.2015.pdf

U.S. DOE/Ventyx. (2010). *Unlocking customer value: The virtual power plant*. http://energy.gov/sites/prod/files/oeprod/DocumentsandMedia/ABB_Attachment.pdf

U.S. DOE/VVO. (2009). *Application of automated controls for voltage and reactive power management.* https://www.smartgrid.gov/files/VVO_Report_-_Final.pdf

WHO. (2016). *Number of registered vehicles.* http://apps.who.int/gho/data/node.main.A995

Ziolek. (2012). *Infrastructure solutions for energy and transport to face upcoming challenges of the future energy supply in Germany.* http://www.nrw.co.jp/file/seminare_jp/20120301/Ziolek.pdf

Chapter 6
Facilitating the Transition Through Digital Technologies

Electricity markets are undergoing a considerable transformation. First, electricity has become the energy of the twenty-first century. Its growth is forecasted to nearly double in the coming 20 years. Electricity is indeed the energy of economic maturity. It is used to power domestic equipment, electronic systems, as well as the large data centers that are becoming dominant in our modern way of life based on information and the use of the Internet. The growth in electricity consumption is first related to the strong evolution of the worldwide population set out for the coming decades. More people connect to the grid and consume more electric energy. It is then related to the economic growth of new economies which, as they develop, require more energy. The energy intensity per capita of these new economies is progressively catching up on the one of mature economies, resulting in a significant increase in consumption worldwide. The growth in electricity consumption is also associated with fuel switching strategies, which privilege the use of electricity to other sources of energy. Traditional biomass consumption in new economies is progressively replaced by electric consumption. As well, electricity often turns out to be one of the most economical ways to consume primary resources of energy. In industry, the yield of electric motors is significantly higher than that of internal combustion engines, leading to industrial enterprises favoring electricity. Finally, in light of global environment issues, transportation itself, which will grow significantly in the coming decades as billions of people in non-OECD countries become more mobile, offers tremendous possibilities for electric transportation, notably in cities. Energy efficiency measures obviously dampen slightly the overall consumption of energy, but not to the extent it changes the upward trend of electricity consumption. Consequently, electricity is set to become the dominant source of energy in the twenty-first century. If the forecasts currently estimate nearly a doubling of the consumption in the coming decades, the overall potential of electricity consumption is beyond limits.

The second element of the electricity markets' transformation is the emergence of competitive renewable energies. Traditionally, power generation has been based on large-scale and centralized power generation units (mostly based on fossil fuels),

© Springer Nature Switzerland AG 2019
V. Petit, *The New World of Utilities*, https://doi.org/10.1007/978-3-030-00187-2_6

which would distribute electricity throughout the grid to millions of users. Renewable energies offer a considerable advantage as there are no greenhouse gas emissions. As well, and maybe most importantly, they do not rely on costly fossil fuel extraction and supply. They also mainly rely on capital costs, which make the investments profitable over time, when conventional generation units see their cost of electricity naturally increase with inflation. The competitiveness of renewable energies has kept increasing in the past few decades. Wind power is already more economical than conventional power generation in many countries, and solar power is set to become competitive everywhere in the coming years. Governments are progressively deploying various environment taxes such as carbon pricing to set a price on the environmental impact of different industries. By doing so, they also mechanically decrease the competitiveness of conventional generation compared to renewable energies. It is thus only a matter of time before renewable electricity actually becomes economically competitive everywhere, leading to a complete change of the power generation mix. Mature economies have already engaged in vast programs of substitution of their current conventional power generation fleet. New economies, which have great needs for additional power generation capacity, more and more privilege renewable energies. The mix of power generation towards renewable is thus set to increase dramatically in the coming decades. In addition, renewable energies present the unique opportunity of a zero marginal cost. This unique specificity will progressively lead to a dramatic change of paradigm. First, because of this specific cost structure, variable renewable energies are expected to take a large portion of base load generation. Moreover, zero marginal cost also means that electrical energy is "free", and the mass deployment of renewable energies could progressively disrupt traditional market structures based on electrical energy (TWh) billing, and consequently the power generation actors of these markets. Similar to what the telecom industry experienced at the beginning of the twenty-first century, 100% of renewable energies would create a free market for energy, where the consumer would only pay for the access to energy. The deployment of capacity markets is a first development in this area. On top of this, free electricity would also lead to massive productivity gains, and probably to a further acceleration of traditional fuel switching to electricity, yielding a significant increase in consumption. These changes will however not come at zero cost. The level of investments in power generation is set to increase in a spectacular manner during the coming two decades, compared to what has been experienced in the past 20 years. Financing is thus the key to ensuring this transition. Here, regulation plays a major role. Governments will most likely not be in a situation to make such levels of investments. Private financing thus becomes paramount, leading to the progressive opening up of markets once public and regulated. Non-OECD markets represent more than 50% of the investments to come. The challenge will be the capability of these countries to open up markets to attract a sufficient volume of financing to ensure the development of infrastructure. In OECD countries, most markets are already accessible to private financing, helping the transition to accelerate. Now, with the relatively stagnant growth of electricity consumption today, these investments in renewable energies will likely be comparatively slow and lead to a restructuring of existing markets.

The progressive penetration of renewable energies leads as well to a number of issues on the grid itself. Traditionally, centralized power generation units connect to a transmission network which then distributes the energy to end consumers through a number of distribution networks. These central generation units are highly controllable. Variable renewable energies, which form the bulk of the new capacities installed worldwide, differ from conventional units. They are by nature distributed, connecting directly to the distribution network, sometimes at the consumer's final point of consumption. Their output is also intermittent due to their dependence on weather conditions. As a result, the higher the penetration of these intermittent sources of power on the distribution network, the more new issues the grid operators have to face. The volume of power capacity to be installed for a given output in energy is higher for renewable energies than for conventional generation units since their capacity factor is lower. Consequently, the volume of power delivered by variable renewable energies can vary significantly at different times of the day. This leads to a rethink of the concept of base loads. The volume of base load generation supplied by conventional technologies will progressively drop, representing a threat to those of these units which cannot easily be turned on and off the grid. Similarly, controllable (thus conventional) power capacities need to be maintained to a certain extent on the grid for those times of the day when variable renewable energies are not able to pick up the demand. A too important penetration of renewable energies on the network can also lead to dynamic and stability control issues, as renewable energies traditionally have no inertia, and therefore a limited ability to instantly react to variations of the parameters of the grid. The lack of predictability of their power output ahead of schedule also leads to an increase in balancing services, which are aimed at maintaining a high enough level of operational reserves to prevent incidents in case the right generation capacity at a certain time to meet the load demand is not available. Flexibility of the system will thus become paramount in the coming decades to mitigate these issues. The development of storage solutions offers in this regards a considerable potential. Going further, because of their distributed nature, renewable energies create alternative flows of energy which the grid has not been designed for, leading to additional and costly grid reinforcement in many parts of the network. The distributed aspect of renewable energies also necessitates a revisit of the traditional architecture of the grid. This second critical characteristic offers the potential to limit to a great extent the cost of distribution of the energy to the consumer point. Self-generation represents a fantastic economical alternative to traditional grid expansion. The development of distributed generation would yield significant load defection, which not only would represent a threat to traditional grid operator revenues, but more importantly lead to the redesigning of network assets. Instead of expensive networks designed to supply full amounts of consumed energies, the networks of the future could be designed as connections across various sources of energy supply, some of them consolidated into microgrids. This new layer of the network architecture could offer significant flexibility and savings on the overall cost of the infrastructure. Traditional grid operators are naturally at the heart of this transformation. Again, this long-term transition will not come at zero cost. During the transition, most issues linked to the growing penetration of renewable energies will require to be financed. Regulators are striving today to find an appropriate answer to mitigate those issues while enabling the transition towards a full renewable based generation fleet.

In summary, renewable energies present over time two major disruptions. On one hand, they offer the promise of "free" electrical energy, likely the most significant productivity disruption of the century. Doing so, they however disrupt traditional market structures. On the other hand, they also present a major potential for savings in the way energy is being supplied, thanks to their distributed nature, leading the overall architecture of grids designed during the first half of the twentieth century to be fully reshaped towards a more decentralized and interconnected organization. Current market structures are unable to sustain this transition efficiently. These changes will however not happen overnight. The necessary transition will also come at a cost. To ensure continuous and reliable energy supply to consumers, existing infrastructure will coexist for a certain period of time with the growing penetration of renewable energies and storage solutions. Market regulators are actively working at reshaping electricity markets in a way that sustain existing actors while encouraging financing of these new infrastructures. In the short term, the key indicator of a successful (and relatively cheap) transition will be the flexibility of operations. Markets will need to evolve towards more real-time, and more geographically accurate operations complemented by flexibility services at all levels of the grid.

Solutions exist to smooth this transition. Smart grids, enabled by digital technologies, are the most promising prospect to enable the transition to occur efficiently. The distribution network is at the heart of this transformation as most renewable energies will be connected to it. Digital technologies alone can help grid operators cope with the increased complexity of the management of the grid, consolidating in real time the various production levels from all these resources and managing the grid in the most efficient and secure way possible. Grid operators face indeed a spectacular disruption as they need more than ever to manage in real time the variations of both load and generation on the grid. The real-time balance issues associated with these variations can only be controlled in a fully digitized network that extends all the way to consumer points. Such a network enables grid operators to understand in real time the issues associated with grid management and suggest the best possible course of action in case of incidents on the grid. Now, the deployment of all these digital technologies leads to a number of issues. The massive influx of data points changes the mission of the grid operator. From managing an infrastructure of cables and switches, it needs to evolve towards managing a very complex infrastructure of data points. This digital transformation leads utilities to rethink their overall IT infrastructure. From a centralized and shared database, the grid operator can develop new applications, such as outage management systems, distribution management systems, variable resources management systems, asset management solutions and workforce management tools. The analytics from such applications will lead the operator to better situational awareness, helping it to control more safely the balance of the grid, to react in a safer and quicker way to incidents, as well as to optimize the efficiency of its assets and workforce, and to better plan the development of the grid infrastructure over time. The grid operator will also be the cornerstone of the development of advanced markets for new actors such as renewable operators, consumers and aggregators, leading to a higher level of grid flexibility and a higher efficiency of the common infrastructure overall. Grid operators are set to become the data manager of the grid of the future.

Independent power producers (IPPs) are also set to rely heavily on digital technologies to better integrate into the grid of the future. Grid codes are expected to become more stringent with regard to the conditions for integrating into the grid, in particular in the area of response to incidents. As well, electricity markets will become increasingly more complex with time, and IPPs will need to better control their generation assets, relying on digital technologies to integrate properly into the grid. They are also expected to invest heavily in aggregation systems which will help them to play efficiently in the various electricity markets with their available generation capabilities in order to maximize their benefits. The development of virtual power plants and advanced market management systems is an example of such evolutions.

Consumers will also be more and more associated with the operation of the grid, as they represent a massive source of virtually untapped flexibility for the real-time balance between generation and consumption. Demand management systems are expected to develop, and new market solutions to emerge, encouraged by the short-term evolution of prices towards more real-time and locational pricing mechanisms. The development of distributed solar energy in residential and commercial buildings is also set to accelerate (as the parity with retail electricity prices will be faster to reach), leading to an acceleration of the development of such mechanisms, as well as the emergence of aggregators which will consolidate these resources (both load and generation) to play actively in the various markets (an avatar of microgrids). Finally, storage technologies represent a formidable prospect. Electricity storage represents the ultimate stage of flexibility, and is therefore the critical enabler of the long-term realization of a new energy system. The current development of batteries (driven by the deployment of electric vehicles) seems to follow a learning curve similar to that of photovoltaic solar modules. This means that the cost of energy storage (combined with distributed generation) is expected to reach a competitive level in the coming decades. With the mass deployment of storage technologies, the market could shift dramatically, accelerating the transition. Massive load defection could then become a reality, and with it the move towards a new model of "free" energy and decentralized grid operations. Final consumers would rely on electricity they would produce (and store) by themselves, rather than relying on that distributed through the grid. It would not lead however to the disappearance of utilities as we know them. Indeed, the lack of seasonal storage and the strong dependability of renewable energies on weather conditions would likely prevent this from happening, especially considering the inelasticity of power demand. Nevertheless, utilities would have to rethink their fundamental role, from distributing energy continuously, meeting demand with offer, to a role where they would provide the capacity to access power at all times, no matter the circumstances, in an interconnected set of microgrids and decentralized generation and consumption. What would be left of the grid and of utility-scale generation would then act as "insurance" for consumers to call upon in case of renewable generation scarcity. Markets would restructure around providing a variety of services issued from the interconnection of these "islands" of generation and consumption. Microgrids would further develop, and with it a new array of intertwined services of generation support and flexibility mechanisms, enabled by distributed (and grid-scale) storage technologies. In this possible future model, digital technologies would play a major role as they would enable the real-time coordination of the different sources of supply (including stored energy) and demand in an interconnected network of millions of islands of energy production and consumption.

Today's electricity markets remain far from this future model, and vary significantly in form from one country to another. This is often due to their history or more simply the conditions of their development. Different countries indeed face different challenges. Some countries face a steep increase in their electricity consumption, while others see their consumption stagnate. Some countries have already a high level of digital technologies deployed on their grid, while others have none. Some countries have already opened their market to private investors, and enjoy a multitude of actors in the market, while other countries (most, actually) remain highly regulated, with verticalized utilities. The variety of situations in different countries is thus part of the complexity of effecting the transition globally. The pathways towards this future will thus be different from one country to another. Nevertheless, it is possible to segment regions at a high level, considering whether or not the transition there is expected to be fast or not, and whether it could be a deep transformation or not. OECD countries in general are expected to experience a deep (if not fast) transformation of their markets. The substitution by renewable energies in those markets is progressive mainly because electricity consumption has remained stagnant (or slightly growing) since the beginning of the century. As a consequence, the penetration of renewable energies more or less comes at the speed of substitution of existing generation units. In countries or regions where the deployment of renewable energies has voluntarily been accelerated (one example is Germany), traditional markets have often required to be heavily restructured. Non-OECD countries are basically split into three groups. Russia, central Asia, and some countries in South America would more likely experience a slow and less significant evolution. They would be the last ones to shift to the new market model. This is due to various reasons: the market is highly regulated and controlled by large public companies; renewable energies will remain uncompetitive for a long time or; like in the case of South America, there is already a high reliance on renewable energies such as hydropower. India, Africa and South East Asia are however expected to experience a fast disruption of their markets. There, electricity demand will skyrocket in the coming decades. The lack of public financing associated with the relative poor development of the infrastructure will likely lead to innovative ideas, such as the development of microgrids. The depth of the changes would thus be stronger there, while the pressure on providing access to electricity to billions of people would make the transition faster. Finally, China and the Middle East are expected to see a fast evolution powered by competitive renewable energies and significant consumption increase, although the market will remain under the control of government bodies and publicly regulated utilities. Every country will thus take a specific pathway towards this new energy landscape, trying to optimize the cost of its transition as well as the critical needs which apply to its infrastructure.

This transition is a fantastic evolution of our time. In the coming decades, thanks to the mass deployment of renewable energies, electricity markets will change radically. 50 years down the road, self-generation could be paramount, and grid infrastructures could have developed as an insurance of power availability, interconnecting a multitude of decentralized generation and consumption sources. Electrical energy could be "free", and consumers would pay to get access to it in a secure and reliable manner. Utilities could consequently have reorganized around flexible markets to optimize in real time the collective capacity infrastructure. With "free" electrical energy, the main productivity step of the century would be passed, leading to new and further improvements of our lifestyle, as well as a more sustainable use of the planet's resources. The transition to a new energy system is already underway all over the world and more and more actors and enterprises are taking part in what promises to be one of the biggest business opportunities of the first half of the century.

The future is with no doubt electric!

Vincent PETIT

December 2017